C. I. Forsyth Major

On Fossil and Recent Lagomorpha

C. I. Forsyth Major

On Fossil and Recent Lagomorpha

ISBN/EAN: 9783741194504

Manufactured in Europe, USA, Canada, Australia, Japa

Cover: Foto ©berggeist007 / pixelio.de

Manufactured and distributed by brebook publishing software
(www.brebook.com)

C. I. Forsyth Major

On Fossil and Recent Lagomorpha

IX. *On Fossil and Recent Lagomorpha.* By C. I. Forsyth Major, *M.D.*
(*Communicated by Prof.* G. B. Howes, *Sec. Linn. Soc.*)

(Plates **36–39.**)

Read 16th June, 1898.

Tooth-change and Tooth-formula in the Lagomyidæ.

THE three extinct Lagomyidæ, *Titanomys*, *Prolagus*, and *Lagopsis*, and the surviving *Lagomys*, have five upper check-teeth, as against six in Leporidæ (*Palæolagus* and *Lepus* s. l.). From a comparison of the form and relative size of the teeth in *Lepus* and *Lagomys*, the type genera of both groups, Waterhouse [*] and Gervais [†] had rightly argued that the last upper molar of *Lagomys* corresponds to the penultimate upper molar in the Hare. Since *Lepus* changes the three anterior of the upper six, and the two anterior of the lower five check-teeth, the formula being therefore $P.\frac{3}{2}, M.\frac{3}{3}$, it might have been further inferred that the number of premolars in Lagomyidæ is the same as in the Leporidæ.

Curiously enough, in recent species of *Lagomys* the tooth-change has never been examined. In 1870 [‡], O. Fraas described and figured the milk-dentition of *Prolagus*, with $\frac{5}{4}$ check-teeth, there being three deciduous molars above and two below. The obvious inference is that the premolars are the same in number as the milk-teeth, and therefore in agreement with what is known in *Lepus*.

Fraas, however, proposes quite a novel definition of what we have to consider to be premolars, with the unavoidable result of thus introducing an element of confusion. Finding the three upper posterior and the three lower posterior check-teeth of *Prolagus* more in agreement as to general form with each other than with those anterior to them, which are two in the upper and one in the lower jaw, he concludes that these last are to be considered as premolars. According to this theory, which conflicts with the prior statement of the number of deciduous teeth, the tooth-formula would be $P.\frac{2}{1}, M.\frac{3}{3}$. But this second statement is again in flagrant contradiction with the following description of the mode in which the tooth-change is supposed to occur. The anterior upper premolar, termed $P._2$ by Fraas, is stated to have no deciduous predecessor, the place of the anterior of the three deciduous teeth being taken by the premolar following behind the first, the so-called $P._1$; while the anterior premolar pierces the jaw in front of $P._1$ and comes in place

---

[*] G. R. Waterhouse, 'A Natural History of the Mammalia,' vol. ii. p. 14 (1848).
[†] Zool. et Pal. Franç., sec. ed. pp. 48, 49 (1859).
[‡] Württemb. naturw. Jahresh. xxvi. p. 169 (1870).

through the same lacuna ("Zahnlücke"), produced by the dropping out of the first deciduous. The two posterior deciduous teeth are, according to the writer, situated on the top of molars I. and II. (!) respectively, like so many caps. So that, according to this description, of the five upper cheek-teeth of *Prolagus*, the first and the last have no deciduous predecessors, but the three intermediate have. In the lower jaw Fraas finds two deciduous cheek-teeth : "Neben dem ersten zweiwurzeligen Deciduus, der über dem einzigen Praemolaren sitzt, ist noch ein zweiter zweiwurzeliger Deciduus, der von dem ersten Molaren verdrängt wird." According to this, in the lower jaw the supposed unique premolar and what he believes to be the first true molar would have deciduous predecessors.

Those astounding views necessarily created a distrust in Fraas' description of $\frac{3}{2}$ deciduous molars (in *Prolagus*); and as a consequence most of the subsequent authors on the subject, up to this day, have, with regard to the Lagomyidæ, preferred to adhere to the old Cuvierian dictum, viz., that in all the Rodents with more than three molars, only the one (or more) anterior to the three are replaced, and that the latter alone are to be considered true molars.

Filhol has observed the two anterior lower cheek-teeth to change in *Titanomys*, and he apparently extends this observation to the maxillary teeth as well : "Chez le *Titanomys*, les deux premières dents étaient sujettes au remplacement" [*].

The one author who first rightly interpreted the tooth-formula of Lagomyidæ is Winge, although he has not seen the tooth-change. Of Fraas' statements he says that they are not clear, partly due to some of the premolars being called molars; and he continues to say that *Lagomys*—which, according to him, includes the fossil "*Myolagus*" and its allies— "has $\frac{5}{5}$ or $\frac{5}{4}$ cheek-teeth; these are the $\frac{2\,3\,4\,5\,6}{3\,4\,5\,6\,7}$ or $\frac{2\,3\,4\,5\,6}{3\,4\,5\,6}$ of the typical $\frac{7}{7}$, as is seen from a comparison with *Lepus*; in the maxillary the three anterior teeth, in the mandible the two anterior are changed" [†].

In the first part of his memoir on Tertiary Rodentia, Schlosser speaks invariably of only one inferior premolar and of a fourth inferior true molar (m. 4) in fossil Lagomyidæ [‡]; but later on he gradually [§] arrives at the true statement of things as given in the supplement to the above memoir, in the following words :—"In this group (*i.e.* the Lagomorpha) at least the first two anterior teeth in each jaw are changed, so that we must speak of two, respectively three premolars" [‖].

My own observations are to the following effect :—

1. *Titanomys*.—This genus has five cheek-teeth in the upper jaw. The deciduous teeth are three in the maxillary and two in the mandible, as is seen in the Rott skeleton described below. The two deciduous inferior teeth, as mentioned above, have already been figured by Filhol [¶].

---

[*] Ann. Sc. Géol. x. p. 29 (1879).

[†] "Om Pattedyrenes Tandskifte" (Vidensk. Meddel. Naturh. Forening i Kjöbenhavn f. 1882), p. 48 (1883). See also H. Winge, in 'E Museo Lundii,' i. pp. 108, 111 (1888).

[‡] 'Palæontographica,' xxxi. p. 10 &c. (1884).

[§] *Op. cit.* p. 110, Anm. 2.

[‖] Palæontogr. xxxi. p. 327 (1885).

[¶] *Op. cit.* p. 29, pl. 3, fig. 3.

As to the number of lower check-teeth, I find, as a rule, five in one of the species, *Titanomys Fontannesi*; but in two out of seventeen mandibular rami there are only four teeth, there being no trace of an alveolus for the last small tooth, which probably will be found constantly present in young specimens.

In the other species, *T. visenoriensis*, the fifth lower molar is supposed to be oftener missing than not. Pomel called *Amphilagus*—regarded by him as a subgenus of *Lagomys*—those specimens of *T. visenoriensis* in which five mandibular check-teeth were present; those with only four teeth he placed in his genus *Lagodus* (*Lagodus picoides*, Pomel, = *Titanomys visenoriensis*, H. v. Mey.). Filhol has based a fusion theory on the presence or absence of the small molar in question *. He assumes that at a certain given moment there prevails a tendency to simplification in the Lagomyine dentition—firstly by the fusion of the last (fifth) tooth with the penultimate, and secondly by the tendency of the fused elements to disappear.

This theory is at once disposed of by the fact that in the mandibles of *Titanomys Fontannesi* before me both the fifth tooth and the posterior colonnette of the fourth—which colonnette Filhol considers to be the fifth tooth fused to the fourth—are present together. I think that for *T. visenoriensis* the same explanation holds good as with regard to *T. Fontannesi*, viz. the fifth tooth has sometimes been lost in the young animal and its alveolus obliterated; its frequent absence is simply explained by the fact that it has dropped out in the fossils.

Anyhow, the formula of *Titanomys* will have to be written as follows:—

$$\text{P.}\ \frac{3}{2},\ \text{M.}\ \frac{2}{2-3},\ \text{or}\ \frac{\text{p.3, p.2, p.1; m.1, m.2}}{\text{p.2, p.1; m.1, m.2 (m.3)}}.$$

2. *Prolagus.*—I have at my disposal the deciduous molars of two species of *Prolagus* [*P. œningensis* (Kön.) and *P. sardus* (Wagn.)]; there are three in the upper and two in the lower jaw, as seen already by Fraas in the first-named species. In the skull of a young *P. sardus*, where the deciduous teeth are *in situ*, the following may be seen :—The anterior of the three deciduous teeth is not situated directly above the anterior premolar, but slightly backward, closely appressed to the second deciduous, so that with its anterior moiety it covers only the posterior part of the premolar; besides it could not possibly cover the latter completely, being much smaller. It is needless to say that neither of the true molars, both of which are already protruded in the skull under observation, supports a milk-tooth; as a matter of fact, the tooth called molar I. by Fraas, which in reality is the posterior of the three premolars, is situated under the posterior of the three deciduous molars, as is the middle premolar under the middle deciduous.

In the lower jaw of both species the two anterior of the four lower check-teeth replace the two deciduous teeth.

Therefore, since *Prolagus* has in the full-grown animal five check-teeth above and four below, its tooth-formula will be:—

$$\text{P.}\ \frac{3}{2},\ \text{M.}\ \frac{2}{2},\ \text{or}\ \frac{\text{p.3, p.2, p.1; m.1, m.2}}{\text{p.2, p.1; m.1, m.2}}.$$

* Ann. Sc. Géol. x. p. 28 (1879).

3 & 4. *Lagopsis* and *Lagomys*.—Since these genera have five cheek-teeth in both jaws, there being a small fifth inferior tooth, their tooth-formula will be:—

$$P. \frac{3}{2}, M. \frac{2}{3}, \text{ or } \frac{p. 3, p. 2, p. 1; m. 1, m. 2}{p. 2, p. 1; m. 1, m. 2, m. 3}.$$

To sum up. The number of premolars is constant in all the genera of Lagomyidæ, and the same as in *Lepus*; whereas that of the true molars varies in the different genera; not *vice versa*, as has been supposed by Lydekker *, Flower †, and Zittel ‡.

The upper m. 3, always present in *Lepus*, is always absent in the Lagomyidæ. Of the lower true molars, m. 3 is always present in *Lagopsis* and *Lagomys*, when not lost in the fossil: it is always absent in *Prolagus*; while in *Titanomys* this tooth is rarely absent in one species, *T. Fontannesi*, more frequently in the other, *T. visenoviensis*, but presumably always present in young specimens of both.

1. Genus TITANOMYS.

Titanomys, H. v. Meyer, Neues Jahrb. 1843, p. 390.

Lagodus, Pomel, Cat. méth. Vert. foss. Loire et Allier, p. 41 (1853); Depéret, Arch. Mus. Lyon, iv. p. 126 (1887).

Lagomys (subg. *Amphilagus*), Pomel, op. cit. p. 42.

Lagomys, Lydekker, Cat. Foss. Mamm. Br. Mus. i. p. 255 (1885).

Lagomys (*Lagopsis*), Schlosser, Pal. Oestr.-Ung. viii. p. 86, foot-n. 4 (1890), p.p.; Depéret ?, Arch. Mus. Lyon, v. p. 58 (1892).

TITANOMYS VISENOVIENSIS.

Titanomys visenoviensis, H. v. Meyer, Neues Jahrb. 1843, p. 390; Gervais, Zool. et Pal. fr., prem. éd., Expl. No. 46, pl. xlvi. fig. 2 (1848–52); Bronn, Leth. Geogn. iii. p. 103 (1853–56); Gervais, Zool. et Pal. fr., deux. éd., p. 50, pl. xlvi. figs. 1, 2 (1859); H. v. Meyer, Palæontogr. xvii. p. 225, pl. xlii. (1870); Filhol, Ann. Sc. Géol. x. p. 26, pl. ii. figs. 25, 26, pl. iii. figs. 1–18 (1879); Schlosser, Palæontogr. xxxi. p. 29, pl. xii. figs. 36, 38, 39, 41, 43, 45, 47, 48 (1884); Zittel, Handb. d. Palæont. i., iv. p. 552 (1891–93).

Titanomys trilobus, Gervais, Zool. et Pal. fr., prem. éd., Expl. No. 46, pl. xlvi. fig. 1 (1848–52).

Lagodus picoides, Pomel, Cat. méth. p. 41 (1853).

Lagomys (subg. *Amphilagus*) *antiquus*, Pomel, op. cit. p. 43.

Amphilagus antiquus, Schlosser, op. cit. p. 30.

Lagomys visenoviensis, Lydekker, Cat. Foss. Mamm. Brit. Mus. i. p. 258 (1885).

Historical Sketch.

In announcing his new genus *Titanomys* (type species *T. visenoviensis*), from the Lower Miocene of Weisenau near Mayence, H. v. Meyer characterizes it as having prismatic cheek-teeth, agreeing in size and number and resembling in form those of *Lagomys*, with the difference, however, that the lower molars of the fossil present a distinct

---

* Cat. Foss. Mamm. Br. Mus. i. p. 255 (1885); Nicholson & Lydekker, Manual of Palæont. ii. p. 1412 (1889).

† Flower & Lydekker, ' Introduct. to the Study of Mammalia,' p. 491 (1891).

‡ Zittel, Handb. d. Palæont. i., iv. p. 551 (1891–93); id. Grundz. d. Palæont. p. 825 (1895).

posterior appendage ("Hinteransatz") not known to exist in *Lagomys*, while the Weisenau Rodent lacks the distinctly developed tooth-particle ("Zahntheil") in the last lower molar of existing *Lagomys* and of those of the "ossiferous breccia"; by which is apparently meant the *Prolagus* of Corsica and Sardinia.

We meet here at the outset with several incorrect statements. The upper molars are not, as we shall see later, prismatic, and the lower are only incompletely so. By the alleged agreement in number of the molars of both *Titanomys* and *Lagomys* we are to understand that both genera have four lower check-teeth, the author believing at that time that the existing *Lagomys* has four mandibular check-teeth, while in reality there are five. H. v. Meyer considered the fifth small cylindric tooth of *Lagomys* to be a third prismatic particle connected with the anterior molar, as is the case in *Prolagus*. The author further makes a distinction—which is repeated two years later in his 'Fossil Mammals of Œningen,' where incidentally the genus *Titanomys* is mentioned *—between a distinct "Hinteransatz" in the posterior molars of *Titanomys*, and the "distinctly developed" posterior or third "Zahntheil" of the last molar in some Lagomyidæ, without being aware that the two are one and the same thing and homologous.

The characteristics given of the upper molars are not incorrect, but rather vague, showing that the author did not succeed in making out the pattern of the triturating surface, as is confirmed also by his manuscript drawings subsequently published by Schlosser.

In the first edition of his 'Zoologie et Paléontologie françaises,' Gervais figures, without description, two mandibular rami from the Lower Miocene of Saint-Gérand-le-Puy (Allier); the fig. 1 of pl. 46 is named *Titanomys trilobus*, the fig. 2 *T. risenoriensis*. In the explanation of the plate it is stated that the identification with *T. risenoriensis* rests on a comparison with a mandible of this species from Germany in the British Museum (this is under No. 21495, from Weisenau). Gervais had no upper molars from the French deposit, but says that those from Germany, which are in London, "sont assez semblables à celles des Lapins, mais beaucoup plus courtes et plus arquées," adding that they are of the same form as those from the Miocene of the Limagne, called *Marcuinomys* by Croizet and *Platyodon* by Bravard. These are two manuscript names.

In 1853 Pomel issued a small work of a high standard on the fossil vertebrates of the Loire and Allier basins, pretending to be nothing more than a catalogue †. The descriptions are in consequence very short, and as there are no figures, the utility of this excellent publication has been rather limited. The Leporidæ family opens ‡ with a new genus, *Lagodus*, from the Tertiary of Langy; the only species, *L. picoides*, scarcely larger than *Lagomys pusillus*, is based mainly on the upper and lower check-dentition, the description of which I transcribe at length for future reference. From this it will be seen that the author assigns to his genus *Lagodus* five upper and four

---

* 'Zur Fauna d. Vorwelt.—Foss. Säugethiere etc. von Œningen,' p. 10 (1845).

† Catal. méthod. et descr. des Vert. foss. découv. dans le Bassin hydrogr. sup. de la Loire, et surtout dans la Vallée de . . . l'Allier (1853).

‡ *Op. cit.* p. 41.

lower cheek-teeth; the first superior was missing, and from the form of the alveolus it is declared to have been very small. "En haut il paraît y avoir eu cinq molaires; la seconde est plus étroite que chez les Lagomys et pour ainsi dire réduite à une seule lame marquée en travers de deux plis d'émail, de manière à figurer presque trois croissants concentriques; les trois autres ont deux lames dont la première est simple, et la seconde pourvue des deux replis d'émail de la dent qui précède, excepté à la cinquième dent, où elle est plus petite.." The lower teeth are said to be four in number, "par absence de la dernière. Première tétragone divisée par deux sillons en deux cylindres comprimés, dont l'antérieure plus saillante est aussi un peu plus large et la seconde a en arrière un petit pli d'émail partant de l'angle interne surtout évident à la dernière molaire et s'effaçant assez tard par la détrition. Ces cylindres sont moins comprimés d'avant en arrière que chez les Lagomys, et leur disque de détrition est ovale oblong, brusquement atténué en angle du côté externe, arrondi vers l'interne."

From the later descriptions of *Titanomys* and from examination of originals, we are enabled to refer Pomel's *Lagodus* to the former genus, and at the same time to appreciate the accuracy of his description. But without this help and in the absence of figures, it becomes difficult to form an exact conception of the complicated pattern of the upper teeth, from their necessarily too short characteristics by Pomel. Hensel, when describing the teeth of *Prolagus* (his *Myolagus*), was on the look-out for allied forms; he gives in full Pomel's description of *Lagodus* *, but fails to see the curious relationship existing between the upper premolars of the former and all the upper cheek-teeth of the latter.

The small enamel fold described by Pomel as starting from the internal angle of the posterior lamina in the three mandibular teeth behind the first is the "Hinteransatz" of H. v. Meyer's *Titanomys*. The relations of the latter to his *Lagodus* are not discussed by Pomel; he suggests the former to be probably the same as *Prolagus sansaniensis* (Lartet's *Lagomys sansaniensis*).

Pomel's *Amphilagus* rests on lower jaws; he considers it to be a subgenus of *Lagomys*, apparently because in both there are five lower cheek-teeth: "la dernière molaire" (in *Amphilagus*) "très petite est cylindrique et caduque, en sorte qu'il ne reste souvent que quatre dents à la mâchoire." The form of the anterior lower premolar is the same as in "*Lagodus*" and *Titanomys*, and very different from the premolar of *Lagomys*, a character which at once suggests that "*Lagodus*" and *Amphilagus* may be identical, and that the absence of the small posterior appendage in the lower molars attributed to *Amphilagus* is due to the specimens being from older individuals than those assigned to "*Lagodus*."

In Bronn's 'Lethæa Geognostica,' Pomel's *Lagodus* is given as a synonym of *Titanomys risenoriensis* on the authority of H. v. Meyer ("*fide* Meyer in litt.").

The second edition of the Zool. et Pal. franç. (1859) gives good reasons for considering *Titanomys trilobus* as the young of *T. risenoriensis*. Of the last lower molar in particular Gervais says:—"la dernière montre encore avec assez d'évidence un troisième lobe, qui

---

* Zeitschr. d. deutsch. geol. Ges. viii. p. 699 (1856).

est d'ailleurs petit et qui, à un âge plus avancé, eût été confondu avec le second lobe de la même dent, comme cela se voit chez le sujet de la figure 2" (*T. risenoviensis*) ; and he goes on stating, as Pomel had done for his *Lagodus*, that this little posterior column is gradually worn away. It is mentioned by Gervais only in the last molar, and his figures show no trace of it in the anterior molars.

Referring to Pomel's *Lagodus* and *Amphilagus*, Gervais launches an ungenerous and unfounded accusation against this author, alleging that the former genus is " du moins en partie " based on his, Gervais', figure of *Titan. trilobus*, and that *Amphilagus* rests on fig. 2, representing *Titan. risenoviensis*. No mention is made of Pomel's description of the upper dentition of " *Lagodus*." If the latter writer failed to recognize in his *Lagodus* and *Amphilagus* H. v. Meyer's *Titanomys risenoviensis*, it was perfectly excusable at the time he wrote, when this species had been so very imperfectly diagnosed both by H. v. Meyer and by Gervais, who both failed to make out the pattern of the upper teeth. Up to this day we have not been better off with regard to the upper cheek-teeth from the type-locality Weisenau.

It would have been fairer on the part of Gervais to acknowledge that Pomel's description of the inferior molars of " *Lagodus* " had gone far in enabling him (Gervais) to recognize the non-validity of his species *T. trilobus*, and that Pomel had besides described more accurately than himself the lower teeth, in demonstrating the presence of the " petit pli d'émail " in *all* the posterior teeth of younger specimens. He certainly could not have based this statement on Gervais' fig. 1 of the young specimen, where only the last molar shows a posterior appendage. The accusation with regard to *Amphilagus* is quite as unfounded as the first one. Pomel assigns five teeth to the lower jaw of his genus, Gervais' figure shows only four ; the description of the first tooth of *Amphilagus* does not exactly agree with the tooth in Gervais' figure, from which last, moreover, it could not be made out that the two cylinders of each of the posterior teeth are united by cement, as stated by Pomel to be the case in his *Amphilagus*. Other particulars occur in the description of *Amphilagus*, which might at once have convinced an impartial critic that Pomel based his description on originals. These were, many years later (1879), handed by M. Pomel himself to Prof. Filhol *.

In his posthumous paper (1870) on the skeleton of a young *Titanomys risenoviensis* from the Lignite of Rott near Bonn, now in the British Museum (No. 41085), H. v. Meyer mentions rooted cheek-teeth in *Titanomys*, and he has been understood to state that only the deciduous teeth of this genus are provided with roots. However, when reading attentively H. v. Meyer's paper—I might almost say, in reading between the lines as well—one necessarily comes to the conclusion that in adult specimens the permanent molars were also rooted, and that the author himself had suspected this fact, but hesitated to proclaim it. Two kinds of rooted *Titanomys*-teeth are mentioned in the paper. With regard to those of the Rott skeleton, the author states that their triturating surfaces are concealed in the matrix, so that their opposite ends only could be examined ; but this does not hide the fact, he continues, that the two posterior upper

* Ann. Sc. Géol. x. pp. 27, 28 (1879).

teeth were formed as in *Lagomys*. This evidently implies that they have no roots; for the writer proceeds to state that in the teeth anterior to those just mentioned lengthened roots can be seen. In the two anterior cheek-teeth of the lower jaw, H. v. Meyer describes a short crown and a long root, composed of two strongly converging parts; and these two teeth seem to be situated somewhat higher than the two posterior, which suggests that they had not yet emerged above the alveolar margin. Contrary to the anterior rooted teeth, these two posterior ones are described as " prismatic"; the whole of their crown has an enamel coating, and is not completely closed below. The author concludes that the teeth seem to indicate that the animal was of immature age, a supposition which would explain the differences of the anterior teeth from those of *Lagomys*.

As a matter of course, in the lagomorphous Rodentia with permanent cheek-teeth growing by persistent pulps, the deciduous teeth are rooted too as in the Rott skeleton. But the author proceeds to state (p. 128) that he has examined detached teeth of the *Titanomys* from Weisenau of two kinds : on the one hand, small teeth corresponding to the anterior teeth of the Rott specimen ; on the other, lower teeth differing from the last by a lengthened prismatic crown and quite insignificant roots; and upper teeth as well, of larger size than those corresponding to the upper anterior teeth from Rott, supposed by H. v. Meyer to be possibly deciduous. In the larger teeth the roots are said to disappear almost completely ; " die flach prismatische, gekrümmte Krone vertritt zugleich die Hauptwurzel, und es wird nur aussen oben ein kleines Würzelchen wahrgenommen, das auch in einer entsprechenden Stelle des Kiefers eingreift, während das an der Innenseite mit einer Rinne versehene Zahnprisma die eigentliche Alveole ausfüllt."

From what will be seen later on. these larger teeth, upper and lower, are in fact the *permanent* teeth of *Titanomys*. as H. v. Meyer hesitatingly suggests. Therefore there is no foundation in the distinction—such as is drawn by Depéret—of two genera, founded on the presence or absence of roots in the permanent teeth, viz. : —

(1) *Titanomys*, with roots in the deciduous set only.

(2) *Lagodus*, with roots in the permanent teeth as well (premolars and true molars).

Proceeding with our historical sketch in chronological order, we next have to consider Filhol's description of *Titanomys risenoriensis* from Saint-Gérand-le-Puy (Allier) *, which has already been quoted more than once in the preceding pages. Among the synonyms of this species are given *Amphilagus antiquus*, Pom., and *Lagodus picoides*, Pom. : the identification of the former rests on one of the type specimens of Pomel; the latter is not discussed in the paper. An important character noted by Filhol is the relatively considerable longitudinal extension of the bony palate in *Titanomys*. The shortness of the bony palate in lagomorphous Rodents is doubtless a specialization ; but by its greater extension *Titanomys* approaches more the condition of other Rodentia and Mammalia generally. The same is true of *Palæolagus*, from the Miocene of North America, which presents curious resemblances with *Titanomys* in its dentition also. Moreover, we meet with a lengthened bony

---

* Ann. Sc. Géol. x. p. 26 (1879).

palate in *Lepus valdarnensis*, Weith. *, from the Upper Pliocene of Tuscany, and in three existing Leporines, *Lepus hispidus*, Pears., from the foot of the Himalayas, *L. Netscheri*, Schleg. & Jent., from Sumatra †, and *Romerolagus Nelsoni*, Merr., from the Popocatepetl (Mexico) ‡, all three of which have other generalized characters in common with each other and partly with *Palæologus*.

Description of Original Specimens.

1. *The Rott Skeleton.*—In its present condition, of the two anterior lower cheek-teeth described and figured by H. v. Meyer, only the imprint is preserved, with the exception of the anterior half of the front tooth, which is still in place. From what can still be seen, and with the help of H. v. Meyer's description and figures, there remains not the slightest doubt that these two anterior teeth belong to the deciduous set, since they bear the characters of milk-teeth, viz., a short crown and (two) long roots, much diverging from each other downward. The number of teeth in front of the two posterior in the upper jaw is left uncertain in the figures and text of the original memoir. A close examination shows that there are three of them : the first apparently is provided with a stouter internal and a somewhat weaker external root; the two following with one internal and two smaller external roots, the latter strongly diverging from the shaft in opposite directions. Here, too, we have the characteristic features of milk-teeth, of which there are consequently three upper in *Titanomys*, as might have been anticipated by analogy to *Prolagus*. The immature condition of the specimen can be further inferred from the fact that the two posterior teeth, viz., the fourth and fifth in the series, are not yet on the same level with the three in front of them. As these two posterior teeth are broken at their lower ends, nothing can be stated as to their roots.

Still less—and this applies to all the teeth of the Rott specimen—can be made out about the pattern of their triturating surface, which, as noticed already by H. v. Meyer, is concealed in the matrix. This deficiency is partly supplied by some teeth from the type-locality of Weisenau, in the British Museum.

2. *Titanomys visenoviensis from Weisenau.*—A fragment of the right upper jaw from the Lower Miocene of Weisenau, in the Geological Department of the British Museum (21495), Pl. 36, fig. 19, shows the two posterior premolars, p.1, p.2, and part of the alveolus of the anterior premolar, p.3. These upper teeth were seen by Gervais, who alludes to them §, contenting himself with the above-reported general remarks. The first of the two premolars preserved, p.2, at once calls to mind by its general form the anterior upper premolar, p.3, of *Lepus*, and to it therefore may be justly applied Gervais' remark referring to all the upper teeth in London, viz., that they are "assez semblables à celles des Lapins." The general outline of this tooth is somewhat triangular, the broader basis being on the inner side, which is imperfectly divided by a slight notch into two abraded

---

* Jahrb. k.-k. geol. Reichsanst. vol. xxxix. p. 80 (1889).

† 'Notes from the Leyden Museum,' vol. ii. note xii. p. 59 (1880).

‡ Proc. Biol. Soc. Washington, x. pp. 169–174 (1896).

§ Zool. et Pal. Franç. 1st ed. t. ii. expl. no. 46 (1848–52) ; 2nd ed. p. 50 (1859).

cusps (8 & 9). Proceeding outward, we meet with two enamel folds starting from the anterior side of the tooth. The one placed more internally (*b*) is by far the larger of the two ; it opens freely on the anterior side, and thence proceeds first internally, and then, gradually attenuating, postero-externally, thus assuming approximately the form of a crescent, whose anterior horn is much shorter than the posterior. Both horns are delimited externally by a cusp (6), having its long axis almost parallel to the long axis of the skull, and protruding with its internal convex border into the enamel fold just described, while its shorter and almost longitudinal external border forms the inner margin of the much smaller second enamel fold (*c*). On the outer side of the tooth we meet with a large bulging enamel tubercle (5), worn by attrition on its inner side only, and showing thus that the outer side in this otherwise much-worn tooth is only partially affected by trituration.

The second tooth, p. 1, presents the general contour of the crown of lagomorphous Rodents, the transverse diameter largely predominating over the longitudinal ; the anterior border is slightly more convex than the posterior. The minute pattern of the triturating surface, however, is very different from that which we are accustomed to consider characteristic of upper leporine molars. The main difference from p. 2 consists in the two enamel folds being shut out from the anterior border by a transverse anterior lobe, which in p. 2 is apparent only in a much reduced condition, its outer portion being entirely wanting. In p. 1 the anterior lobe or "wall" delimits the anterior horn of the enamel fold (*b*) on its front side, so that in this tooth the anterior horn is much more lengthened transversely than the posterior. As compared with p. 2, p. 1 has undergone, as it were, a lateral pressure, by which the various parts of the surface have been forced into a more transverse direction. This is apparent, especially in the strong cusp (6) separating enamel folds *b* and *c*, which is no more longitudinally directed as in p. 2 but has likewise assumed the form of a crescent with its convexity projecting inward into enamel fold *b*, and forming externally the inner margin of enamel fold *c*. The latter has in its turn assumed a more transverse direction, and is only incompletely shut out from the outer border of the tooth by a blunt enamel tubercle (5), occupying mainly the postero-external part of the tooth. The summit only of this tubercle is slightly worn.

The inner border of p. 1 is more distinctly divided than in p. 2 into two abraded cusps by a vertical groove, manifesting itself on the triturating surface in the shape of a short enamel fold, or notch (*a*).

The levelling effect of trituration—favoured by the enamel folds in both teeth being more or less completely filled with cement,—together with the more transverse direction assumed by the folds and cusps of p. 1, tends to produce a lophodont character of its triturating surface ; or rather, we have a selenodont type leaning towards lophodonty.

3. *Tilanomys visenoviensis*, from the Allier (France). Br. Mus. 31094 (Bravard Collection).—A detached tooth (Pl. 36, fig. 12) is more worn still than that just described, as revealed by its triturating surface being more flattened and the enamel folds more narrowed. It can only be either p. 1 or m. 1. P. 3 is quite out of the question, as, to judge from its alveolus, it was a very small tooth ; p. 2 is reduced in its antero-external,

m. 2 in its postero-external portion (compare fig. 19, Pl. 36, with fig. 6); so that the choice remains only between p. 1 and m. 1. It resembles closely the p. 1 described; only it is narrower, and the anterior lobe of the latter is more developed in its internal portion, although the inverse was to be expected, the p. 1 described being less worn. For these reasons I think it more likely to be m. 1. This tooth shows two small roots on the outer side; on the inner side the crown gradually thins out downward into a single large root. It cannot be a milk-tooth, because the two small external roots do not diverge downwards, but run parallel with each other. We have here another proof, if one were needed, that in *Titanomys visenoviensis* the upper teeth are provided with roots; although this fact has been denied with regard to this species of the Lower Miocene.

Mandibular teeth of Titanomys visenoviensis. As a characteristic feature of the lower check-teeth of *T. visenoviensis*, H. v. Meyer makes mention of a small posterior lobe, calling it a distinct posterior appendage ("ein deutlicher Hinteransatz") [*]. About the occurrence of this small particle much uncertainty prevails. When establishing the genus, in the paper just quoted, H. v. Meyer mentioned it in a general way as present in the lower check-teeth, seemingly implying that all of them were provided with this appendage. In his posthumous memoir, however, speaking again of the Weisenau specimens, he says that it occurs on the *posterior* check-teeth and would have disappeared by effect of attrition [†]. Pomel assigns it to the three posterior check-teeth of "*Lagodus picoides*," adding that it takes its origin from the internal angle, that it is more evident especially in the last molar and disappears rather late by attrition [‡].

According to Gervais [§] it would occur only on the fourth tooth (m. 2), and as a character of young specimens; the same is stated by Filhol [||], who had at his disposal a considerable number of lower jaws. Schlosser [⌐] styles it a third lobe occurring as an anomaly ("abnorm vorkommend") in "m. 3" (meaning m. 2) of *Titanomys visenoviensis*; although in the same memoir he figures manuscript drawings by H. v. Meyer, where it is shown in two molars. This same small lobe occurs in *Palæolagus* also; it is transitional in one species, *P. Haydeni*, as described by Cope [**], apparently persisting in another species, *P. tripler* [††]. On a former occasion I incidentally pointed out the interest attached to it from both a phylo- and ontogenetic point of view [‡‡].

As to the occurrence of this small lobe or cusp in *T. visenoviensis*, my own observations tend to show that it is constantly present in young specimens, not only of the posterior, but also of the anterior lower teeth, including p. 2. In a fragment of a right mandibular ramus of *T. visenoviensis* from the Allier (Bravard Collection, Br. Mus. 31094–104), Pl. 37, fig. 25, exhibiting the two anterior check-teeth, p. 1 and p. 2, in a moderate stage of wear, traces of this lobe are visible in both these premolars, very distinctly in the posterior (p. 1).

* Neues Jahrb. 1843, p. 390.

† Palæontogr. xvii. p. 226 (1870).

‡ Cat. méth. p. 41 (1853).

§ Zool. et Pal. Franç. sec. ed. p. 50 (1859).

|| Ann. Sc. Géol. x. p. 27 (1879).

⌐ Palæontograph. xxxi. p. 32 (1884).

** ·The Vertebrata of the Tertiary Formations of the West,' p. 876 (1883).

†† Op. cit. p. 881.

‡‡ Proc. Zool. Soc. London, p. 203 (1893).

It has been found convenient to give the detailed descriptions of the lower molars of this and all the other genera in a separate chapter (p. 473).

TITANOMYS FONTANNESI.

Lagodus Fontannesi, Depéret, Arch. Mus. Lyon, iv. p. 127, pl. xiii. figs. 19–19 *c* (1887).

Lagomys (Lagopsis) verus, Schlosser, Pal. Oestr.-Ung. viii. p. 86 (1890) ; Depéret (?), Arch. Mus. Lyon, v. p. 57 (1892).

Under the name of *Lagodus Fontannesi*, Depéret described a fragment of an upper jaw, from the Middle Miocene of La Grive-Saint-Alban (Isère), as related to *Titanomys visenoviensis*, H. v. Meyer; but, in addition to its larger size, he distinguished it by other more important characters.

Schlosser has supposed, without assigning reasons, that *Lagodus Fontannesi*, Dep., is synonymous with *Lagomys (Lagopsis) verus*, Hens. (= *Lagomys œningensis*, H. v. Mey.), and Depéret, in his second publication on the Fauna of La Grive, is disposed to accept Schlosser's views. It may be asked at once, what then becomes of the left palate, figured and described by Depéret in his first memoir *, where he considers it, rightly in my opinion, to belong to the *Lagomys verus*. As this question will be discussed under the head of *Lagopsis verus*, when it will be shown that Depéret's original view in distinguishing between " *Lagodus Fontannesi*" and *Lagomys verus* is the correct one, we have for the present only to deal with Depéret's first memoir, in which " *Lagodus Fontannesi*" is described, and where he asserts that it is distinct as a genus from *Titanomys visenoviensis* of the Lower Miocene.

For this Depéret gives two reasons. In the first line he maintains that his *Lagodus* preserves in its adult dentition part of the characters of the deciduous dentition of *Titanomys visenoviensis*, meaning that in the latter the milk-teeth alone are rooted, while in the former the permanent cheek-teeth are rooted as well. I have already disposed of this supposed difference, by showing that the permanent teeth of *Titanomys visenoviensis* are likewise rooted.

Depéret's second reason is given in the following words:—" Le *Lagodus Fontannesi* se distingue d'ailleurs facilement du *Titanomys visenoviensis* ... par quelques différences dans les dessins d'émail qui ornent la surface de la couronne " (*i. e.* of the upper molars). "D'après M. Filhol, le lobe postérieur des molaires supérieures du *Titanomys* d'Auvergne est orné d'un double pli en chevron entourant une pointe externe ; dans le *Lagodus* de La Grive il y a trois plis en chevron concentriques et pas de pointe extérieure bien manifeste "†.

The enlarged figures of the triturating surface in the teeth of " *Lagodus Fontannesi*" and *Titanomys visenoviensis* do not help us, as they are sadly inaccurate. The artist who drew the former ‡ completely failed to understand the pattern ; while in Filhol's enlarged drawings§ the artist has not even made an attempt at accuracy, contenting himself with drawing the outlines of the teeth, and leaving out almost completely the details of the

---

* Arch. Mus. Lyon, iv. p. 164, pl. xiii. fig. 17 (1887).

† *Op. cit.* p. 128.　　　　　‡ *Op. cit.* pl. xiii. fig. 19 *b*.　　　　　§ *Op. cit.* pl. iii. fig. 15.

crown's surface. In the figures which I give of the teeth of both forms *, no essential difference is to be seen in the pattern. The disagreement in the description of the two writers finds its explanation in the somewhat loose way of describing the triturating surface, *i. e.* by the failure to distinguish between a dentine surface bordered by two enamel ridges which alternates with an enamel fold filled with cement, so that only the two enamel borders of the fold appear on the surface. As an outcome of this alternation we find, when proceeding from the inner side of the tooth to its outer side, the following succession in the middle line of the tooth : enamel ridge ; dentine ; enamel ridge ; cement ; enamel ridge ; dentine ; enamel ridge ; cement ; enamel tubercle of the outer side.

Apparently the two writers do not always apply the term "chevron" to the same thing. Filhol, speaking of the "deuxième élément" of the tooth, by which he means the part of the crown backward from the anterior lobe, his "premier élément," says : " Chez les *Titanomys*, on peut le considérer comme constitué par un chevron à sommet interne, dont les deux extrémités circonscrivent une pointe externe. Ce premier chevron est borné en dedans par un deuxième chevron dont le sommet correspond au bord interne de la dent. Ce mode de structure est surtout bien marqué sur la troisième molaire"†. This description, which is quite correct as far as it goes, applies equally well to the species of the Lower and to that of the Middle Miocene, as may be seen by a comparison of the figures (Pls. 36, 37, 39) ; by consulting the figures it may be further seen that what the author calls chevrons are the spaces of dentine bordered by enamel ridges, which spaces mark the position of enamel cusps before wear set in.

Depéret, in describing the same "troisième molaire," *i. e.* the posterior of the three premolars, of *Lagodus Fontannesi*, says :—" Cette couronne se compose de deux prismes d'émail étroitement accolés, un peu mieux distincts en dehors que du côté interne, qui est de forme arrondie. Le prisme antérieur [Filhol's premier élément] est composé d'un seul pli d'émail transverse ; le prisme postérieur au contraire, à surface triturante coupée obliquement en arrière, présente deux plis d'émail en chevron à pointe interne, ce qui dessine sur la couronne trois petits croissants concentriques, si l'on compte la lamelle d'émail qui limite le bord interne de la couronne "‡.

It is certainly not accurate to describe the single cylinder of which these upper teeth consist as composed of two enamel prisms "étroitement accolés." Apart from this, Depéret's description, like Filhol's, applies to both *Lagodus Fontannesi* and *Titanomys visenoviensis*. By " deux plis d'émail en chevron à pointe interne," the author evidently has in view, firstly, the larger, internal, of the two enamel folds ; secondly, the crescent-shaped cusp (6) external to it, which by the effect of wear presents a dentinal surface bordered by an outer and an inner enamel ridge. By counting, moreover, the enamel border of the internal side of the crown, Depéret arrives at the number of three " petits croissants concentriques," which on the following page are called " trois plis en chevron concentriques." Filhol leaves out of account the enamel fold by which his two chevrons are separated.

---

* Pl. 36. figs. 18, 19 ; Pl. 37. fig. 11 ; Pl. 39. fig. 16 (*Titanomys visenoviensis*). Pl. 36. figs. 6–8, 12–15 (*T. Fontannesi*).　　　† *Op. cit.* p. 30.　　　‡ *Op. cit.* p. 127.

As a result of this minute analysis we find that there exists no essential difference in the tooth-pattern of the two supposed genera.

The roots of the *Titanomys*-molars have next to be described. I have elsewhere dealt incidentally with the conditions in *Titanomys* (*Lagodus*)*. I was impressed by the fact that the chief points of wear are on the inner side in the upper, on the outer in the lower molars, and that these parts are the first to appear lengthened (vertically) in teeth in a condition intermediate between brachyodonty and hypselodonty, while the outer sides of upper, and the inner sides of lower molars remain, as it were, in a passive condition (for upper molars of *Titanomys* see Pl. 39. figs. 1, 2, 5, 13, 14, 19). It then appeared to me that the upper teeth of *Titanomys* showed the hypselodonty—which, as above demonstrated, is here in fact "accompanied by a gradual and essential change of the pattern of the crown"†—to extend gradually towards the outer side. In the description of the pattern of the *Prolagus*-molars (pp. 452, 453) I have reconsidered my former view, and have been able to show that the obliteration of the original pattern is chiefly the consequence of an atrophy on the outer side; whereas the secondary pattern is brought about by a new addition, starting from the inner side and directed chiefly *inward*. It remains none the less true "that the vertical elevation of the crown, the first stage towards hypsodonty, always has its starting-point from the *inner* side of upper molars"‡. I added at the same time that "the inner root (of the upper molars) which ultimately will remain open, gradually extends outward, increasing in size, and receives a coating of enamel"§. It is against this latter assertion especially that the Rév. Père Heude has directed a criticism, couched in energetic terms ‖. When he begins by saying that I had not demonstrated my assertion, he is perfectly right; but I had at the time no other intention than to assert, reserving full demonstration for a work on the Lagomorpha under preparation, as intimated on p. 208.

The Rév. Père's arguments are to the effect that the roots of teeth cannot be imagined to receive a coating of enamel, because brachyodonty "est un arrêt de développement, une fixation par cessation de mouvement, une détérioration du fût transformé en racine. Conséquemment la dent ne peut revenir à son mouvement initial." In order to demonstrate that "*logiquement*" hypselodonty is more primitive ("plus ancien") than brachyodonty, and that "réellement ces deux faits sont phylogéniquement indépendants," the Rév. Père adduces the incisors of Rodentia. "D'autre part toutes les incisives des Rongeurs étant essentiellement hypsodontes et à toutes les époques, au point qu'elles emportent la définition de l'ordre, il faut admettre qu'elles n'ont pas varié, qu'elles ont un caractère commun fixé, et qu'à ce titre l'hypsodontisme est plus général que le brachyodontisme."

It is not hypselodonty, as such, which is the more primitive condition, but the growing of a tooth by a persistent pulp. And, since hypselodont teeth continue to grow by persistent pulps during the greater part or the whole of the animal's life,

* Proc. Zool. Soc. London, 1893, p. 206.　　　　† *Ib.*　　　　‡ *Ib.*　　　　§ *L. c.*

‖ 'Mémoires concernant l'Hist. naturelle de l'Empire Chinois, par des Pères de la Compagnie de Jésus,' t. iv. p. 75 (1898).

they may, in a sense, be termed primitive; but, as a matter of course, brachyodont and semihypselodont teeth, before they are perfectly developed, have the cavities at their bases open as well as hypselodont teeth; and when they are in this condition, their brachyodonty is not yet "un arrêt de développement." Ontogenetically and "logiquement," every hypselodont tooth passes through a brachyodont condition, the shaft only gradually increasing in length. Phylogenetically, brachyodonty is also more primitive than hypselodonty, as is known to all scientific morphologists who have a knowledge of palæontology.

On Pl. 39. figs. 19 and 20, I have delineated side by side in the anterior view a posterior upper right premolar, p. 1, of *Titanomys Fontannesi*—the same specimen of which the upper view is figured on Pl. 36. fig. 8—and an upper right molar of a young *Pteromys*, in which the roots are not yet closed. Fig. 14 represents the anterior view of a right upper molar of *Tit. cisenoriensis*, figured in upper view on Pl. 36. fig. 18. Now, if we are entitled to call roots, even though they be imperfectly developed, the three prolongations of the crown in *Pteromys* (fig. 20), I think we are justified in applying the same term to the evidently homologous parts in the figured teeth of *Titanomys* (*cf.* figs. 14 and 19, and figs. 1, 4, 5, and 13), and in repeating what I have said formerly [*], that the inner root of *Titanomys*, which ultimately will remain open, increases in size and receives a coating of enamel.

Even perfectly adult brachyodont teeth preserve at their extremity a minute opening for the passage of nerves and vessels, so that it may be left to individual judgment at which phase in the ontogeny or phylogeny of a tooth we may begin to use the term "root." Having no desire to juggle with words I would, be quite ready to desist using this term for the part of the tooth of *Titanomys* which is the homologue of the inner root of *Pteromys*; but thereby nothing would be altered. The question at issue is, whether or not a coating of enamel has extended to that part; and that this has been the case is shown plainly enough by the figures.

It is interesting to compare the tooth of *Tit. cisenoriensis* (fig. 14) with those of *Tit. Fontannesi* (figs. 1, 13, 19). The small outer roots are perfectly closed in the former and more detached from the shaft than in the latter. The tooth of the former, as shown by the upper view (Pl. 36. fig. 18), is from an old individual; but in none of the numerous upper premolars or molars of *Tit. Fontannesi* have I met with closed outer roots. The coating of the enamel does not extend so far downward on the inner side in *Tit. cisenoriensis* as in *Tit. Fontannesi*.

A further difference between the Lower and the Middle Miocene species is also characteristic. In the former (Pl. 39. fig. 14) the external part of the crown extends more outward than in *Tit. Fontannesi*, beyond the small roots; this character has been already noticed and explained in the description of the triturating surface, as due to the atrophy of the outer region being less advanced in *Tit. cisenoriensis* than in the more recent species.

To proceed now to a closer examination of the small outer roots of the upper molars and premolars of *Titanomys*. In a passage, quoted above, p. 440, from H. v.

Meyer's posthumous description of the Rott skeleton, mention is made of upper molars of *Titanomys* found isolated, but only one small outer root is ascribed to them. I likewise find that the anterior milk-tooth, d. 3, of the Rott skeleton has one small outer root. Almost all the isolated teeth at my disposal, of both species, exhibit two symmetrical outer rootlets, which represent the lower free terminations of two prominent ridges on the upper outer region of the tooth, as in the figured deciduous tooth of *Lepus* Pl. 39. fig. 9, b), with the difference that in the latter the posterior part of the first appears higher, and the ridges, therefore, more lengthened than in *Titanomys*. The ridges, of which the outer rootlets are the lower terminations, are present also in molars and premolars of all Lagomorpha growing from persistent pulps. Figs. 7 and 8 (Pl. 39), representing germs of the first upper true molar of a rabbit, show them in side view (at the right side of the figures).

In a left upper jaw of *Titanomys Fontannesi* the roots of the cheek-teeth are described in the following manner by Depéret:—"La disposition des racines est aussi très particulière, et diffère de ce que l'on voit chez les Léporidés pour se rapprocher d'autres groupes de Rongeurs tels que les Spermophiles. Chacune des quatre dernières molaires porte trois racines, dont une interne grosse, ovalaire transversalement, et deux externes relativement très petites et arrondies. L'alvéole de la première molaire est petit et rond : il annonce une molaire uniradiculée et à couronne assez petite "*. The figure of the specimen † shows the empty alveoli of p. 2 and m. 1, so that the mode of disposition of the roots in the jaw can be seen. Depéret's description is confirmed and supplemented by the figure which I give (Pl. 36. fig. 23) of a left maxillary from which the teeth have dropped out.

P. 2 of *Tit. risenociensis*, the anterior lobe of which we have seen to be somewhat reduced antero-externally (Pl. 36. fig. 19), as compared with the posterior teeth, has only one outer rootlet (Pl. 39. fig. 5 a) ; in the place of the antero-external rootlet it displays a curious conformation, which gives at once a clue to that of the rootless molars of the other lagomorphous genera, and explains why the upper teeth described by H. v. Meyer have one outer rootlet only. There is no free antero-external radicle to this tooth ; but, as seen in the side-view (fig. 5 a, Pl. 39), a raised ridge runs along its antero-external side down to the bottom, where, as shown in the lower view of the tooth (fig. 5, b), it is confluent with the lower opening of the large inner root, the homologue of the widely open cavity in the genera (*Lagopsis*, *Prolagus*, *Lagomys*, *Lepus*) with rootless teeth.

To judge from its alveolus, p. 2 of *Titanomys Fontannesi* was more like p. 1 and the true molars, than p. 2 of *Tit. risenociensis*.

Fig. 2, Pl. 39. represents (a) the anterior, and (b) the outer view, of the last upper molar, right side, of *Tit. Fontannesi*, the upper view of which has been figured in Pl. 36. fig. 6. Both outer rootlets are broken off, but they seem to have had a free

---

* Charles Depéret. "Rech. sur la Succession des Faunes de Vert. Miocènes, etc." Arch. Mus. Hist. Nat. Lyon, t. iv. p. 171 (1887).

† *Op. cit.* pl. xiii. fig. 19.

termination like the teeth anterior to them (figs. 1, 13, 19). The outer view (b) shows the whole of the outer side devoid of enamel.

The levelling effect of trituration tends to produce a more lophodont character of the crown. In an unworn condition, however, these teeth present a much more bunodont appearance, and it requires a very small effort of imagination to trace them back—conspicuously so the intermediate in the series, which are more typical—to a more brachyodont as well as bunodont form, in which the predominant feature is that the cusps, while the intervening enamel folds would appear as shallow valleys, are not yet filled with cement. We meet with such brachyodont types in the Eocene (classed as Creodonts and Lemuroids); more than any other, the Eocene " *Pelycodus helveticus* Rüt.," and *Plesiadapis*, both so-called Lemuroids, show teeth in close agreement with *Titanomys*. Let, *vice versâ*, a brachyodont molar of the shape of " *Pelycodus, helveticus* " (Pl. 36. fig. 3) or *Plesiadapis* (Pl. 36. fig. 2) become somewhat more hypselodont by the heightening of its shaft, and let the valleys between the cusps be filled with cement, and the result will be a *Titanomys*-tooth. This I had in view when, on a former occasion *, I stated that the structure of the lagomorphine molar can be traced back to a " pelycodoid type."

2. Genus PROLAGUS.

Lagomys, G. Cuvier, Oss. foss. iv. pp. 21, 22 (1812), sec. ed. iv. pp. 200, 203 (1823) ; Rud. Wagner, Kastner's Arch. f. d. ges. Naturlehre, xv. pp. 14, 18 (1828) ; id. Oken's Isis, p. 1136 (1829) ; p. p. H. v. Meyer, Neues Jahrb. 1836, p. 58 ; p. p. id. Foss. Säugeth. etc. von (Eningen, p. 6 (1845) ; Waterhouse, Nat. Hist. Mammalia, ii. p. 32 (1848); Lartet, Not. Colline de Sansan, p. 21 (1851) ; p. p. Fraas, Württ. naturw. Jahresb. xxvi. p. 171 (1870); Lydekker, Cat. Foss. Mamm. Brit. Mus. i. pp. 256, 257 (1885), v. p. 325 (1887).

Anoema, König, Icones Foss. Sectiles, pl. x. fig. 126 (1825).

Prolagus, Pomel, Cat. méth. p. 43 (1853).

Myolagus, Hensel, Zeitschr. deutsch. geol. Ges. viii. p. 695 (1856).

Archaeomys, Fraas, Württ. naturw. Jahresh. xviii. p. 130 (1862).

G. Cuvier was first to recognize that some fossil remains, which belong to the above genus, are those of a lagomorphine Rodent ; he figured and described them from an ossiferous breccia of Corsica, and later from a breccia of Sardinia, considering them to be a species of *Lagomys*.

In 1825 König figured, in his ' Icones Foss. Sectiles,' a skeleton from (Eningen.

H. v. Meyer (1836) notes among the Mammals of (Eningen the genus *Lagomys*; the same, according to Murchison, had been previously suggested by Laurillard †. H. v. Meyer further supposes that König's *Anoema* might belong as well to the former genus.

From the Miocene of Sansan (Gers) and Venerque (Haute-Garonne), Lartet mentions a lagomorphine Rodent of the size of a large rat, which he proposes to unite with

---

* P. Z. S. 1893, p. 208.

† R. I. Murchison, " On a Fossil Fox found at (Eningen, near Constance," Trans. Geol. Soc. London, iii. 2, p. 285 (1832).

Lagomys, on account of its having one superior molar less than the genus *Lepus*; adding, however, that the Sansan fossil has one inferior molar less than the existing *Lagomys*.

For this same Rodent from Sansan, Pomel proposed to create the sub-genus *Prolagus*, on the ground of its differing from *Lagomys* " par la dernière molaire inférieure, qui a trois prismes par réunion de la cinquième molaire à la quatrième. Du reste, la première est aussi triangulaire. On pourrait nommer l'espèce *Prolagus sansaniensis*." The hypothetical fusion of two molars, stated here as an undoubted fact, does not stand close investigation, any more than in the case of *Titanomys*. But to this I shall return in the sequel.

An excellent description of the remains of the lagomorphine Rodent from the ossiferous breccia of Sardinia is given by Hensel. He founds on them his new genus *Myolagus*, and points out that one of the two Lagomyidae from Œningen, *Lagomys Meyeri*, v. Tschudi, is closely related to the Sardinian fossil, and therefore likewise to be placed in the genus *Myolagus*. (It is a pity that the perfectly well-characterized *Myolagus* has, for priority's sake, to give way to Pomel's " *Prolagus*," just as it is to be regretted that Pomel's amply-described *Lagodus* has to stand back before H. v. Meyer's imperfectly characterized *Titanomys*.) Hensel refers to Pomel's *Prolagus* *, and rightly observes that the characters mentioned by the latter writer recall to mind the genus *Myolagus*; he considers them, however, to be insufficient for a decision. This was quite true at the time when Hensel wrote. It is incorrect to say, as has been done by H. v. Meyer†, that Hensel based his genus uniquely on the form and number of the lower cheek-teeth and the position of a foramen mentale. Hensel had laid great stress also on the pattern of the upper teeth ‡, a character which H. v. Meyer, as in the case of *Titanomys*, studiously avoids discussing.

A step backward is made by Fraas, when he figures and describes a well-preserved mandibular ramus from Steinheim under the name of *Archæomys steinheimensis*. He was set right by H. v. Meyer §, who referred the supposed *Archæomys* from Steinheim to " *Lagomys* (*Myolagus*) *Meyeri*, Tschudi," and in 1870 he atoned for his mistake by giving a full description of the Steinheim Rodent in question.

PROLAGUS ŒNINGENSIS.

Anœma œningensis, König, Icones Foss. Sect. pl. x. fig. 126 (1825).

Lagomys œningensis, p. p. H. v. Meyer, Neu. Jahrb. p. 58 (1836).

Lagomys œningensis, Waterhouse, Nat. Hist. Mammalia, ii. p. 32 (1848).

Lagomys Meyeri, v. Tschudi, in H. v. Meyer, Zur Fauna d. Vorwelt. Foss. Säugeth. etc. von Œningen, p. 6, pl. ii. figs. 2, 3, pl. iii. fig. 2 (1845); Lydekker, Cat. Foss. Mamm. Brit. Mus. i. p. 257 (1885).

Lagomys sansaniensis, Lartet, Not. Coll. de Sansan, p. 21 (1851).

Prolagus sansaniensis, Pomel, Cat. méth. p. 43 (1853).

---

* Op. cit. p. 702.　　　　　　† Palæontogr. xvii. p. 228 (1870).

‡ Op. cit. p. 895.　　　　　　§ Neu. Jahrb. 1854, p. 197; 1865, p. 843.

Myolagus Meyeri, Hensel, Zeitschr. deutsch. geol. Ges. viii. p. 699 (1856) ; Fraas, Württ. naturw. Jahresh.
    xxvi. p. 171, pl. v. figs. 2–16 (1870) ; Schlosser, Palæontogr. xxxi. p. 28, pl. xii. fig. 44 (1884).
Archæomys steinheimensis, Fraas, Württ. Naturw. Jahresh. xviii. p. 130, pl. ii. fig. 19 (1862).
Lagomys (*Myolagus*) *Meyeri*, H. v. Meyer, Neu. Jahrb. p. 197 (1864), p. 843 (1865).
Lagomys verus, p. p. Fraas, Württ. Naturw. Jahresh. xxvi. p. 171 (1870).
Prolagus Meyeri, Depéret, Arch. Mus. Lyon. iv. p. 123. pl. xiii. figs. 18–18 *c* (1887).
Myolagus sausaniensis, Filhol, Ann. Scienc. géol. xxi. p. 46, pl. i. fig. 8 (1891).
Lagomys (*Prolagus*) *Meyeri*, Depéret, Arch. Mus. Lyon, v. p. 55, pl. i. figs. 30, 31 (1892).

When publishing his first note on *Lagomys*-like Rodents from Œningen (1836), H. v.
Meyer was not aware that two rather different forms occur there; he comprises them
both under the name of *Lagomys œningensis*. Later on, in his Monograph of the fossil
Vertebrata from Œningen, he arbitrarily sets aside König's specific name for the smaller
form, for which he adopts a manuscript name by v. Tschudi, *Lagomys Meyeri*, found on
one of the labels, while he reserves the name *Lagomys œningensis* for the larger
form. As stated before, the same author identified the lagomorphine Rodent from
Steinheim with the smaller form from Œningen; and in the sequel equally those from
several other Miocene deposits in Germany.

On the ground of Pomel's description of the Sansan species, Schlosser adds *Lagomys*
(*Prolagus*) *sausaniensis*, Pomel, to the synonyms of *Myolagus Meyeri*; and likewise the
Lagomyidæ from the Spitzberg in the Ries, near Nördlingen (Bavaria), referred to
Lagomys verus, Hens., by Fraas (1870). Filhol has figured as *Myolagus sausaniensis*
(E. Lartet) the type-specimen, a mandibular ramus, of Lartet's *Lagomys sausaniensis*,
and is satisfied that "cette espèce, comme on le verra par l'examen de la figure grossie
que nous en donnons, était très différente de toutes celles qui ont été décrites" [*]. It is
precisely this enlarged figure of the lower cheek-teeth which shows conclusively that the
Sansan fossil is one and the same with the *Prolagus* species from Œningen and Steinheim,
as conjectured by Schlosser and confirmed by Depéret [†], who has added La Grive-
Saint-Alban (Isère) [‡], Mont-Ceindre, and Gray [§] to the localities of this widespread
Middle Miocene species.

The following descriptions are based on specimens collected at La Grive-Saint-Alban
by myself.

In the genus *Prolagus* the molars are no longer rooted, and, with the exception of the
deciduous teeth, all the cheek-teeth grow from persistent pulps. It does not, however,
follow that the triturating surface preserves throughout the animal's life the same
pattern. This is the usually accepted belief [‖]; but although the proofs to the contrary

---

[*] Ann. Sc. Géol. xxi. p. 47, pl. i. fig. 8 (1891).

[†] Arch. Mus. Lyon, v. p. 57 (1892).

[‡] *Op. cit.* iv. p. 167 (1887), v. p. 56 (1894).

[§] *Op. cit.* v. p. 57.

[‖] See, *e. g.*, Giebel, in Bronn's 'Klassen und Ordnungen des Thierreichs,' vi. v. p. 152 (1875), where he treats of
the Rodentia with laminated teeth (" Blätterzähne "), including the Lagomorpha. He says of them : " Die
Kauflächen dieser Zähne ändern ihre Zeichnung durch Abnutzung nicht." He might have known better, at least
as regards the Hares, from what Hilgendorf had said ten years before (Monatsber. K. Preuss. Akad. d. Wissensch.
Berlin, 14 Dez. 1865, p. 673) respecting the upper grinding-teeth of young Hares.

are not in all cases so evident, and so surprising at the same time, as in the group under consideration. or as in Geomyidæ *, or *Haplodontia* †, it is nevertheless a fact that neither in Rodents nor in Mammalia generally is the surface of the crown absolutely identical throughout its length ; although in many of them we may speak of a *relative* constancy of pattern.

Hensel, in the description of the upper teeth of *Prolagus*, has overlooked this circumstance, and as a result has in one case wrongly interpreted the tooth-structure. This occurs in the description of " *Myolagus sardus* ; " ‡ but, since Fraas has endorsed Hensel's error in his description of *Prolagus œningensis* (Kön.) (" *Myolagus Meyeri* ") §, which differs very little from the former, we shall have to deal with the argument in the present description as well. How little both Hensel and Fraas were aware of the change of pattern depending on the age of the animal is shown by the way in which, for convenience sake, they studied the tooth-crown. Hensel does not figure the natural surface of attrition, but gives transverse sections of it ‖; while Fraas declares ¶ that it is more convenient to examine the teeth from the inferior side. meaning the open alveolar end of the shaft !

Fig. 21, Pl. 36, represents the four upper grinding-teeth of *Prolagus œningensis* in a rather worn condition. Both the upper true molars, the fourth and fifth in the series, those teeth which in *Titanomys* exhibit a beginning of reduction on the postero-external side, have undergone in *Prolagus œningensis* a considerable change as compared with the same teeth in the former genus. Of the two more or less crescentic enamel folds of *Titanomys*, only one, apparently the inner, persists, in the form of a very small enamel islet in the posterior part of the triturating surface (*b*). The notch of the internal side (*a*) has been transformed into a transverse enamel fold, which, as we shall find to be likewise the case in *Lagomys* and *Lepus*, approaches the outer side of the tooth. The enamel lining of the outer side, partially interrupted in the postero-external corner of m. 2 of *Titanomys*, is almost entirely missing in the external border of both the molars of *Prolagus* (and of its posterior premolar as well). In other words, the outer parts of the crown, those which are the least affected by trituration, have degenerated in consequence of disuse ; and we might be inclined to assume that compensation has been effected by the transverse fold penetrating towards the outer part. But this is not, to all appearance, the exact explanation of the phenomenon. The triturating surface in the tooth of the young animal—in the part of the shaft which is the earliest formed—is more square than in the adult ; in the latter, it presents the well-known narrow transverse shape of the lagomorphine upper molar. If we remove one of these teeth from its socket and examine it from the anterior or posterior side, it can be seen that,

---

* C. Hart Merriam, 'Monographic Revision of the Pocket Gophers. Family Geomyidæ' (North American Fauna, no. 8), pl. 16 (1895).

† Proc. Zool. Soc. London. p. 706 (1897).

‡ Zeitschr. deutsch. geol. Ges. viii. pp. 690, 691 (1856).

§ Württ. naturw. Jahresh., xxvi. pp. 174, 175 (1870).

‖ " Die Backenzähne sind stets senkrecht zu ihrer Axe angeschliffen worden, daher sind die Abbildungen eigentlich eine Aneinanderreihung der einzelnen Querschnitte " (*l. c.* p. 703).

¶ *Op. cit.* p. 173.

while its outer border maintains throughout its height almost a vertical direction, or is even concave, the inner part of the tooth slopes down medially, from below to above (taking into account its natural position in the maxillary). The tooth, therefore, as it continues to grow, extends persistently in a transverse direction; but this growth takes place chiefly, if not exclusively, towards the internal side; so that the transformation of the internal notch of the *Titanomys*-tooth into the transverse fold of the true molars of *Prolagus* is not the result of its extension outward but inward. In other words, new formation takes place in that part of the tooth where there is increased work, while the outer part—that which is scarcely or not at all affected by trituration—not only remains stationary, but even becomes atrophic.

On the other hand, since in the more specialized forms, beginning with *Prolagus sardus* (Pl. **36**. fig. 24), the transverse enamel fold reaches almost the outer side in the true molars, it is very possible, and even likely, that secondarily a slight extension *outward* of this transverse fold takes place; although the outer border of the tooth is nearly functionless, its condition, almost devoid of enamel, would nevertheless effect a too rapid wear of the dentine if some compensation were not ensured.

The posterior of the three premolars, p. 1, situated between the first molar and the second premolar, is intermediate in shape as in position. Both the crescentic enamel folds of *Titanomys* are preserved in this tooth in the shape of central enamel islets, a much larger internal one (*b*), with an elongate anterior horn, and a smaller outer one (*c*) (fig. 21). The transverse fold (*a*) which opens on the inner side is much shorter than in the molars; it is scarcely more than an elongate notch. To put it otherwise, as compared with the molars, p. 1 presents less *reduction* in its external part, and less new formation in regard to the transverse fold starting from the inner side. Exactly the inverse is apparent when we compare p. 1 with the tooth in front of it.

This latter (p. 2) has triangular contours, with the apex internally, a shorter, slightly convex anterior, and a longer posterior side; as a consequence, its outer border runs obliquely. In its pattern, this premolar strongly resembles the *Titanomys*-teeth; instead of a transverse enamel fold we find in it a short notch (*a*) on the inner side, as in *Titanomys*; while almost the whole of the crown-surface is occupied by the two crescentic enamel folds (*b* and *c*), with an indication of a minute third one—equally marked in *Titanomys*—on the antero-external corner. The latter is more distinct in younger specimens of *Prolagus œningensis* (Pl. **36**. fig. 10, p. 2). The enamel folds alternate with crescent-shaped, pointed cusps.

On comparison of p. 2 with p. 1 it becomes at once clear that the main difference between the two consists in the circumstance that the crescentic enamel folds in the former have become reduced to the condition of enamel islets, their communication with the antero-external margin of the tooth having ceased. When describing p. 1 of *Prolagus sardus*, in which, as a comparison of our figures shows, this tooth (fig. 24, p. 1) is almost identical with its homologue in *P. œningensis*, Hensel labours under a strange misconception. He says:—"Das Merkwürdigste aber an dem Zahn sind zwei isolierte Schmelzcylinder. Sie befinden sich in dem äusseren und hinteren Viertel des Zahnes." [*]

* *Op. cit.* p. 690.

After describing these enamel cylinders accurately, he continues :—" Der Inhalt dieser beiden kleinen Cylinder ist ganz gewiss Zahnbein, obgleich eine mikroskopische Untersuchung nicht angestellt werden konnte. Man sieht aber an dem Wurzelende des Zahnes die beiden Cylinder, sowie den ganzen Zahncylinder, hohl, daher sie auch wie dieser sich später wohl mit Zahnbein füllen werden. Wir haben hier ein Beispiel einer Zahnbildung, die bisher noch nicht beobachtet wurde. Denn hier ist nicht eine Vereinigung einzelner Cylinder zu einem Ganzen wie bei den sogenannten zusammengesetzten Zähnen, sondern eine Einschachtelung "[inclusion]" zweier einzelner Zähnchen in einen grossen." *

It seems strange that so accurate an observer should not have perceived at once that the islets ("isolierte Schmelzcylinder") of p. 1 are the homologues of the two enamel folds which, on the preceding page, he had described in the anterior tooth (p. 2); and that an enamel fold whose central part dips vertically, and deeper in the shaft of the tooth than the peripheral, generally becomes by attrition reduced to a central islet This is a phenomenon of the most common occurrence in teeth of all Mammalian orders. Hensel's misconception is intelligible only from his apparently not being aware that teeth growing from a persistent pulp, like the brachyodont teeth, though only to a certain extent, are liable to changes in the pattern of their triturating surface.

As a matter of course the enamel islets of p. 1 are filled with cement, as are the enamel folds of the anterior tooth. The argument adduced by Hensel goes for nothing, as not only the dentine, but also the cement is always missing in the root-ends of these teeth, both substances being not yet developed in these younger stages.

As mentioned on a preceding page. Fraas has endorsed Hensel's statements, when describing the similar-fashioned p. 1 of *P. aningensis*. He is, besides, of opinion that the deciduous teeth furnish the explanation of the conformation of p. 1 :—" Die Betrachtung der Milchzähne wirft auf diese in der That von allen bekannten Zähnen abweichende Bildungsweise ein Licht." † A supposed extraordinary phenomenon calls for an extraordinary explanation, and this he gives when describing the deciduous teeth ‡. He means to say that there is a connection between the roots and the enamel folds, inasmuch as the cylindrical roots are included in (or by) the tube composing the whole tooth, as it were, nested in it (" eingeschachtelt ")—just as we should speak of willow-boxes nested one into the other—the folds appearing on the surface of attrition, according to this theory, being but the upper ends of the cylindrical roots! The only thing which the author thinks remarkable is the fact that the central folds, which are in connection with the roots, are present as well in the permanent teeth which are devoid of roots. At the bottom of this singular theory lies, first, the author's initial statement, to the

---

* *Op. cit.* p. 691.   † *Op. cit.* p. 175.

‡ *Op. cit.* pp. 177, 178. "Die Falten....die auf der Kaufläche des Zahns zu Tage treten sind nichts anderes, als die oberen Enden der in die Zahnbüchse eingeschachtelten Wurzelcylinder. Sehen wir somit an den Milchbackenzähnen auf deren Oberfläche Schmelzfalten zu Tage treten, wo die Wurzeln sich vereinigen, so begreift sich dieser Faltenschlag leicht. Das Eigenthümliche ist nur, dass sich die inneren, mit den Wurzeln zusammenhängenden Falten auch an den permanenten Zähnen zeigen, die aber ihre ganze Dauer wurzellos sind. Es ist diess, so zu sagen, die Uebertragung eines Jugendzustandes auf das Alter....' etc.

effect that the roots of the deciduous teeth of *Prolagus* have a coating of enamel : " sie
bestehen genau aus derselben Schmelzmasse, wie die Zahnbüchse selbst, die das Zahnbein
umgiebt " * ; and secondly the fact that in some instances he seems to have mistaken
for roots what in reality are the tube-like lower terminations of the enamel folds.

In the first tooth of the upper series (p. 3, Pl. **36**. fig. 21) the two enamel folds are
also present ; they penetrate into the surface of the crown from its anterior side
and run in a longitudinal direction. The anterior border (" wall ") of the triturating
surface, already slightly shortened in p. 2, is still more shortened in p. 3, being reduced
to a short longitudinal stump on the antero-internal corner.

From what has previously been stated, we are prepared to find, in different stages of
attrition of these upper teeth, some difference in the pattern ; this is, in fact, what takes
place. The enamel islet of m. 2 has disappeared in old specimens ; and such is the case in
the specimen figured by Fraas †. The enamel islet of m. 1 varies in size according to
age, being larger in younger specimens. The same holds good with regard to the two
enamel islets of p. 1. We anticipated that in young stages of this tooth the enamel
islets would have the shape of enamel folds opening freely on the margin of the tooth,
as is the case in p. 2. This is, in fact, what happens in young specimens of the following
species (*P. sardus*). Of *P. œningensis* I have no very young examples.

P. 2 varies little with age; the notch on the inner side is more distinct in comparatively
young individuals, and there is shown in this stage (fig. 10) a third very small enamel
fold in the postero-external corner of the tooth, which soon disappears by attrition.

Deciduous upper teeth of P. œningensis.—Fraas has figured the three deciduous upper
cheek-teeth *in situ* ‡ ; he scarcely describes their pattern, contenting himself with the
statement that the anterior one is well provided with folds (" faltenreich "), and that it
presents much resemblance to the *second* of the permanent dentition §.

I have only detached upper deciduous teeth, five in number. Two of these are in the
British Museum, under M5237, from my collections. The anterior milk-tooth (d. 3) is
not represented among these five detached teeth ; according to the figure of Fraas, and
to what I know of the same tooth of *P. sardus*, it has triangular contours ; while the
detached teeth at my disposal are squarish oblong, almost tetragonous, their transverse
diameter slightly exceeding the longitudinal. They show (Pl. **36**. fig. 29) an internal
notch and two enamel folds, the latter opening freely on the outer side. The internal of
the two folds (*b*) has the form of a crescent and is the larger of the two. The roots
are three in number; the outer two very minute, the inner single one considerably
larger ; the former run parallel with each other, but not with the odd inner root, which
strongly diverges from them inward, while they diverge outward (Pl. **39**. figs. 21, 22).

PROLAGUS SARDUS.

Lagomys sardus fossilis, Rud. Wagner, Oken's Isis, p. 1156 (1829),.
Lagomys fossilis, Id. op. cit. p. 1139.

---

* In this there is some truth ; see above, pp. 446, 447.
† *Op. cit.* pl. ii. fig. 6. ‡ *Op. cit.* pl. ii. fig. 14. § P. 177.

Lagomys corsicanus, Rud. Wagner, op. cit. p. 1139; Giebel, Fauna d. Vorwelt, i. p. 99 (1847); Gervais, Zool. et Pal. franç., first ed. p. 32 (1848), second ed. p. 50 (1859); Lortet, Arch. Mus. Lyon, i. p. 53, pl. 8 (1873).

Myolagus sardus, Heusel, Zeitschr. deutsch. geol. Ges. viii. p. 695, pl. xvi. figs. 7, 8, 11 (1856); Forsyth Major, Atti Soc. Ital. Milano, xv. p. 390 (1873); id. Kosmos, vii. (vol. xiii.) pp. 6, 7 (1883).

Lagomys (Myolagus) sardus, Schlosser, Palæontogr. xxxi. p. 29 (1884).

Lagomys sardus, Lydekker, Cat. Foss. Mamm. Brit. Mus. i. p. 256 (1885), v. p. 325 (1887); Schlosser, Pal. Oestr.-Ung. viii. p. 86 (1890).

This Pleistocene species, which is somewhat larger than its Middle Miocene forerunner, closely resembles the latter in its upper molars, as the comparison of the figure shows. However, the specialization of the true molars has progressed, for in the teeth of the adult no trace remains of the two crescentic enamel folds (Pl. 36. fig. 24). P. 1 agrees in the two species. P. 2 is scarcely different in either; the enamel folds in p. 2 of the adult *Prolagus sardus* are slightly reduced in size, and the larger inner fold (*b*) is, in old specimens, sometimes shut out from the outer border by intervening dentine (fig. 24, p. 2). P. 3 has its anterior "wall" somewhat more developed than in *Prolagus œningensis*.

Of this species I have collected a good number of young specimens. The examination of younger stages of the teeth is of considerable interest, as they recall, more than the adult teeth, the primitive features of the *Titanomys*-type.

Firstly as to p. 2. This tooth, being the most conservative, shows, as might have been anticipated, the least change from young to old. The diminutive postero-external enamel fold, however, which we met with in a moderately young specimen of *P. œningensis*, is visible only in very young individuals of *Prolagus sardus*.

P. 1, as has been intimated above, exhibits in the young stage a close approach to p. 2: the two enamel folds are not yet reduced to the shape of islets, but open freely on the outer side of the tooth (Pl. 36. fig. 11); the only appreciable difference, apart from its square outline, consisting in this, that the crescentic cusp (6) which divides the two enamel folds has its anterior horn less produced outward, so that the folds unite in a common delta on the outer side. The next stage of the still young p. 1 (fig. 16) is the pattern we met with in old p. 2; the small external enamel fold (*c*) alone opens on the outer side, while the larger internal fold has been reduced to the shape of a crescentic islet (*b*). The third stage is that of the adult, the external fold likewise having become an islet (fig. 24).

It might be expected that very old specimens of p. 1 would show the complete disappearance of the islets, as is the case in the true molars; this condition I have never found in *Prolagus sardus*, although I have had the opportunity of examining more than a hundred upper jaws. But it occurs in a Pliocene form of Continental France (Roussillon), of which more will be said hereafter.

In the youngest stages of the anterior true molar (fig). 4, in jaws which still preserve the deciduous dentition, remains of the two enamel folds are still visible; they are very imperfectly divided by the last trace of the once powerful intermediate cusp. In a slightly more advanced stage (Pl. 36. fig. 16 (m. 1), one or two diminutive enamel islets, situated postero-externally to the internal end of the transverse fold, are the last vestiges of the enamel folds of m. 1. In rare cases, very young m. 2 likewise show at the same place a diminutive circular enamel islet, fig. 16 (m. 2).

The deciduous teeth (Pl. **36**. fig. 4) are scarcely different from those of the preceding species; but in these teeth also the crescentic cusp "6" does not completely divide the two enamel folds. D. 3 is triangular; d. 2 in younger stages somewhat approaches to a triangular contour.

PROLAGUS LOXODUS (Gerv.).

Lepus sp., Gervais, Zool. et Pal. Fr. 1ʳ éd. i. p. 32 (1848).

Lepus loxodus, Gervais, ib. ii. explic. pl. xxii., pl. xxii. fig. 9 (1848–52).

Lagomys loxodus, Gervais, Zool. et Pal. Fr. 2ᵉ éd. p. 50 (1859) ; id. Zool. et Pal. gén. p. 148 (1867–69).

? *Lagomys (Prolagus) corsicanus*, Depéret. Mém. Soc. Géol. France, i. p. 56 (1890), iii. p. 122, pl. xii. figs. 1, 1 *a* (1892).

? *Myolagus elsanus*, Forsyth Major, Atti Soc. Tosc. Sc. Nat. i. p. 229, 238 (1875), &c. (*vide infra*).

Gervais' *Lagomys loxodus* has been a stumbling-block for fifty years, owing, as I think, to the circumstance that the pattern of the four posterior right upper check-teeth preserved had not been grasped and was incorrectly represented. An inspection of the original specimen would at once settle the question; but since I am not acquainted with the original, I must deal, as best I can, with the published figure and Gervais' incomplete description.

The figure is four times natural size. Gervais' description runs as follows:—"Diffère des Lagomys actuels et diluviens par la forme ovalaire et sublosangique des doubles lobes de ses seconde à quatrième molaires supérieures ; la molaire antérieure est en même temps plus forte, et elle a ses replis plus compliqués ;—taille sensiblement inférieure à celle du Lapin de Garenne"*. It was found in the town of Montpellier, in the fluviatile Pliocene marls †. At the same locality, under the Palais de Justice, was found the *Semnopithecus monspessulanus*; and this circumstance is of importance, as proving that these fossils belong to the older of the two faunas, mixed together under the designation Montpellier. *Semnopithecus* occurs also in the Lower Pliocene of Casino (Tuscany).

The reason for which Gervais considered the teeth to be the first, second, third, and fourth is obvious ; the last in the series is equal in shape to the penultimate, while in *Lagomys* the last molar has a postero-internal appendage. I believe them to be the second, third, fourth, and fifth (p. 2, p. 1, m. 1, m. 2) of a species of *Prolagus*, because the anterior tooth has the characteristic triangular outline of p. 2 of *Prolagus*, with the apex turned inward (*cf.* pl. **36**. figs. 10, 21, 24, p. 2). In further agreement with *Prolagus*, Gervais' figure of this tooth exhibits on the outer side two enamel loops; on the inner, one. The more minute features of this tooth, as well as of those following behind, were not recognized, and therefore the latter have been represented in the manner in which lagomorphous upper teeth generally were and still are, founded on the belief that they are composed of two distinct lamellæ closely connected.

In my opinion there is not the slightest doubt left that we have to do with a species

---

* Zool. et Pal. Fr. 2ᵉ éd. p. 50.   † L. c.

of *Prolagus*, and I feel sure that a close examination of the fossil, if it still exists in the Museum of Montpellier or elsewhere, will confirm my view.

It remains to enquire whether there is some reason for identifying it with one of the species of *Prolagus* found in deposits contemporaneous, or approximately so, with the strata of Montpellier in question. Of these there are two: (1) *Prolagus* (*Myolagus*) *elsanus*, which I have mentioned from the lignites of Casino, in the Val d'Elsa, Tuscany; and (2) " *Lagomys* (*Prolagus*) *corsicanus*," described under this name from Roussillon by Depéret[*]. The little I have to say of the former will be stated in a separate paragraph hereafter.

As to the latter, Depéret declares that it agrees in size as well as in all other characters with the *Prolagus* from Corsica and Sardinia, and he therefore describes it under the above name. This proceeding is as it should be; so long as no differences are traceable between both there is no reason for two specific names. But, so far as my own experience goes, the circumstance of a mammalian species surviving unaltered from the Lower Pliocene to the present era (I have found calcified remains of *Prolagus sardus*, var. *corsicanus*, in an "abri sous roche" of the Neolithic period in Corsica) would be quite unique, and it is *a priori* highly improbable, even taking into consideration that insular species may become, up to a certain extent, conservative in their character. I therefore incline towards the belief that hereafter characters distinguishing the Roussillon from the island form will be shown to exist.

The presence of a third lower molar, supposed by Depéret to appear occasionally in the Roussillon fossil, would be such a distinctive character, since it has never been observed in the Pleistocene species; but I give further on (pp. 482, 483) what I hold to be the real explanation of the fact noticed by Depéret, viz. that the supposed m. 3 in certain jaws from Roussillon is simply a portion of m. 2, which has been accidentally detached.

Another character noticed by Depéret in the Roussillon species deserves mention here. In the specimen from this locality first described [†] it was stated that the three posterior upper cheek-teeth are similar to each other, being "construites sur le type ordinaire des Léporidés." In the third volume of the 'Mémoires' a second specimen is described [‡]; in this the "première arrière-molaire" (p. 1) differs from the same tooth of the first specimen by "exhibiting on the surface of its posterior lobe a double chevron-shaped enamel fold, recalling the molars of *Titanomys*. These folds must disappear rather rapidly by effect of trituration, thus explaining their absence on the specimen previously figured, which apparently was more adult." Depéret adds that these chevron-like folds exist equally in the corresponding tooth in the specimens of " *Lagomys corsicanus* " from Bastia (Corsica), although this character is not represented in the figure of the latter published by Lortet [§], and he concludes that the above is a complete confirmation

[*] Ch. Depéret, " Animaux pliocènes du Roussillon," Mém. Soc. Géol. France, i. p. 56, pl. iv. figs. 27–35 (1890); iii. p. 122, pl. xii. figs. 1, 1 *a* (1892).

[†] Mém. Soc. Géol. France, i. p. 57 (1890).

[‡] *Op. cit.* iii. p. 122, pl. xii. figs. 1, 1 *a* (1892).

[§] Arch Mus. Lyon, i. pl. viii.

of the identity of the Corsican and Sardinian fossil with the Pliocene animal from Roussillon.

I venture to suggest that the inverse conclusion may be drawn from these statements. The character in question has been figured and exhaustively described in the preceding pages. Of *Prolagus sardus*, I have represented on Pl. 36. three stages. Fig. 11 shows p. 1 of a young individual in which the two enamel foldings (*b* and *c*) are large and confluent on the outer margin. In fig. 24 (p. 1 from an adult and rather old individual), they are seen to be separated from each other and reduced to the shape of central enamel islets. Fig. 16 exhibits an intermediate condition (see p. 456). If these chevrons are not represented in Lortet's figure quoted by Depéret, this is due to an inadvertence of the artist; for an examination of the figure quoted shows that the artist had seen something of the kind, but omitted to represent it accurately. In the vast number of maxillaries of *Prolagus sardus* from Bastia and various Sardinian localities which have passed through my hands, I have never missed the presence in p. 1 of the two enamel folds: but it is possible that they may disappear in very old individuals. The fact that, of the only two specimens from Roussillon examined, this character was absent in one, proves in my opinion that the Roussillon species, although geologically older, has exceeded the island species in the transformation of the cheek-teeth, thus representing the last stage of *Prolagus*; *i. e.* that which approaches closest to the condition shown by p. 1 of *Lagopsis* and *Lagomys*.

The peculiarity which I am about to mention in the anterior lower premolar of the *Prolagus* from Casino is not recorded by Depéret in the lower p. 2 from Roussillon; but it would be worth while to re-examine this tooth in the specimens from the latter place: for the two *Prolagi* from Roussillon and Casino may be identical, if we judge from the association of other identical species in the two localities. The same may be said of the fossils from Roussillon and Montpellier; but the information concerning the *Prolagus* from the latter locality at present at our disposal is insufficient for close comparison with other fossil forms.

PROLAGUS ELSANUS, Maj.

Myolagus elsanus, Forsyth Major, Atti Soc. Tosc. Sc. Nat. i. pp. 229, 238 (1875); id. in L. Rütimeyer, Ueber Pliocen und Eisperiode auf beiden Seiten der Alpen, p. 15 (1876); id. Atti Soc. Tosc. Sc. Nat. Proc. Verb. p. xc, 11 Maggio 1879.

A few fragmentary mandibular rami from the Lower Pliocene lignites of Casino, Val d'Elsa (Tuscany), preserved in the Pisa Palæontological Museum, have been long ago noticed by me, and I have on various occasions stated that, by the conformation of their lower anterior premolar (p. 2), their reference to Hensel's genus *Myolagus* (*Prolagus*) is beyond doubt. As at the time no species of *Prolagus* had been recorded from the Lower Pliocene, I felt justified in assigning a new specific name to the Casino fossil.

Of some importance, not only as distinctive for the species, is the following character not previously recorded by me, but of which I was perfectly aware at the time, for it is shown in two sketches which I made of the lower anterior premolar, right and left, presumably of the same specimen. At the postero-internal margin of this p. 2 is a

64*

narrow enamel fold—more distinct in the left-hand tooth—corresponding to a shallower and wider fold in *Titanomys*, which forms the anterior boundary of a minute terminal cusp, marked "*t*" in the figures (*Titanomys*, Pl. **37**. figs. 2, 3, 7, 25). More about the significance and the homologies of this terminal cusp will be said in the chapter treating of the lower check-teeth. I mention it here, since in no other species of *Prolagus* have I met with it in p. 2, and it may therefore be characteristic of *Prolagus elsanus*.

The only teeth known from Casino are mandibular; and as those from Montpellier are maxillary, no direct comparison can be made between them. Both deposits are contemporaneous, and bear other species in common; wherefore there are good grounds for assuming the specific identity of the remains of *Prolagus* from the Italian with those of the French deposit. If this can be satisfactorily shown in the sequel, Gervais' specific name will have to replace mine on grounds of priority.

3. Genus LAGOPSIS, Schloss.

LAGOPSIS VERUS (Hensel).

Lagomys œningensis, H. v. Meyer, Neu. Jahrb. 1836, p. 58, p. p. ; id. Foss. Säugethiere &c. von Œningen, p. 6, pl. iii. fig. 1 (1845) ; Biedermann, Petrefacten aus d. Umgeg. v. Winterthur : H. Die Braunkohlen von Elgg, p. 13, pl iii. figs. 1, 2, 3 (1863) ; Lydekker, Cat. Foss. Mamm. Brit. Mus. i. p. 256 (Specim. Br. Mus. nos. 42815, 42816 (?), 42820 (?) (1885).

Lagomys verus, Hensel, Zeitschr. deutsch. geol. Ges. viii. p. 688, pl. xvi. figs. 12, 13 (1856) ; Depéret, Arch. Mus. Lyon, iv. p. 161, pl. xiii. figs. 16, 17 (1887).

Titanomys œningensis, H. v. Meyer, Palæontogr. xvii. p. 228 (1870), p. p.

Lagomys (Lagopsis) œningensis, Schlosser, Palæontogr. xxxi. p. 31 (1884), p. p.

Lagomys (Lagopsis) verus, Schlosser, op. cit. p. 31, pl. xii. figs. 10, 46, 49 (1884) ; Depéret, Arch. Mus. Lyon, v. p. 57 (1892), p. p.

Hensel's type-specimen is a mandibular ramus, and will therefore be more fully discussed in a later chapter. He was impressed by its approaching much nearer the recent *Lagomys* than the remains of *Prolagus* ("*Myolagus*") described in the same paper. " Ich nenne die Art *Lagomys verus*, weil sie sich durch die Zahl ihrer fünf Backenzähne, durch die Stellung des Foramen mentale und durch den ersten unteren Backenzahn, der nur aus einem Cylinder besteht, als ein ächter *Lagomys* ausweist " *.

It is perfectly true that this fossil is closely related to *Lagomys*. However, Schlosser proposed to raise " *Lagomys œningensis*, H. v. Mey.," and " *Lagomys verus*, Hens.," to the rank of a genus, *Lagopsis*, a position with which I in general agree, while I disagree in part with the reasons assigned for it. There is no doubt that some of the larger Lagomyidæ of Œningen, which were comprised by H. v. Meyer under the above name, are identical with Hensel's *Lagomys verus*; but with regard to other specimens this has not yet been demonstrated. We cannot therefore throughout regard " *Lagomys œningensis*, H. v. Mey.," as synonymous with " *Lagomys verus*, Hens.," as Schlosser has hesitatingly assumed in his ' Nager des europ. Tertiärs ' (p. 32) and more positively asserted later †, followed by Lydekker ‡.

---

* Op. cit. pp. 688, 689.   † Beitr. Pal. Oestr.-Ung. viii. p. 86 (table) (1890).
‡ Cat. Foss. Mamm. Brit. Mus. i. p. 256 (1885).

Schlosser bases his new genus *Lagopsis* on the differences (from *Lagomys*) in the shape of the anterior lower premolar (p. 2), " und das, wie es scheint, häufige Fehlen des vierten Molaren," thereby meaning the lower m. 3. I agree with the first proposition ; as to the latter, it will be shown later on that in all the specimens of *Lagomys verus*, in which m. 3 is missing, it has simply dropped out, for its alveolus is present.

The upper teeth of *Lagopsis*, which are here described for the first time, although more closely resembling *Lagomys* than *Prolagus*, present, however, characters which strengthen the conclusion based on the lower teeth, viz. the establishment of a separate genus. *Lagopsis* realizes the penultimate stage in the evolution of the cheek-teeth of Lagomyidæ, *Lagomys* the last.

The description of the upper cheek-teeth of *Lagopsis* may be appropriately preceded by that of *Lagomys* [*]. The numerous existing species of *Lagomys* show a considerable constancy in the pattern of their cheek-teeth. Young individuals were not available to me. In the adult we find a further step away from the *Titanomys* type ; not only the two true molars, but the posterior premolar (p. 1) likewise, have lost every trace of the crescentic enamel folds, so that p. 1 has become very similar to the true molars, all three showing the transverse fold proceeding far outward. P. 2 exhibits, in a very interesting manner, a reduction of the *Titanomys* type. There is no transverse fold, the original internal notch being maintained ; of the two crescentic enamel folds (*b*) and (*c*) only the former, the internal, remains, and it bears on its outer side a strong cusp (6) and opens on the antero-external margin of the tooth. P. 3 shows a further reduction as compared with *Prolagus*. Of the internal notch only a feeble vestige is visible, and of the two typical enamel folds only the internal one, which runs obliquely from the middle of the anterior margin in a postero-external direction.

Depéret has figured from La Grive a left palate devoid of teeth, but exhibiting very distinctly the alveoli of the five cheek-teeth ; he assigns this fossil, quite rightly in my opinion, to *Lagopsis verus* [†].

Among the fossils collected by myself at La Grive are two rooted upper cheek-teeth (Brit. Mus., G. D., No. 5264), which in size agree with the lower teeth of *Lagopsis verus* from the same deposit. *Lagopsis* being the one Tertiary genus which, by the form of its lower teeth, comes nearest to *Lagomys*, it could be anticipated that the upper teeth of the fossil would likewise show a near approach to the recent genus, and this is, in fact, the case. Additional evidence is furnished by a specimen from Œningen, to be described later on.

One of the isolated teeth just mentioned, from La Grive (Pl. 36. fig. 31), exhibits the same somewhat triangular outline—the apex being turned outward—and about the same characteristic enamel folding (*b*) as the upper p. 2 of *Lagomys*. In the p. 2 of *Lagomys* the outer enamel border of the crescent (*b*) is raised into a strong triangular cusp, with its convexity turned inward ; in the fossil tooth the inner border of the crescent is raised in the same manner. From p. 2 of *Prolagus œningensis* (Kön.) (Pl. 36. fig. 21) the tooth

---

[*] Enlarged horizontal sections of the upper cheek-teeth of *Lagomys alpinus* and *L. nepalensis* are given by Hensel. *op. cit.* pl. xvi. figs. 1 & 5.

[†] *Op. cit.* p. 164, pl. xiii. fig. 17.

figured in fig. 31 can at once be distinguished; the former is much smaller, has a tri-angular outline with the apex turned inward, and a smaller enamel crescent (*c*), smaller than, and external to (*b*). The upper teeth of *Titanomys Fontannesi*, which in size come nearer to the original of fig. 31, though slightly smaller, are provided with roots, and they present other differences which have already been described. From its resemblance to *Lagomys* this tooth (fig. 31) can therefore with certainty be determined as belonging to *Lagopsis verus*. The second of the isolated teeth before mentioned, from La Grive (Pl. 36. fig. 32), agrees in size with the first; and for this reason alone *Prolagus œningensis* can be excluded. It is either p. 1 or m. 1, if we judge from its agreement with the corresponding teeth in *Lagomys*.

In the Palæontological Collection of the British Museum (No. 42815) is preserved a slab from Œningen, showing the skeleton, " in a much crushed and imperfect condition," of a lagomyid Rodent, which Lydekker has determined as *Lagomys œningensis*, H. v. Mey., because it agrees very closely in size with that figured by H. v. Meyer on pl. iii. fig. 1 of his ' Fossile Säugethiere von Œningen ' *. On examination of this specimen (No. 42815) several cheek-teeth are seen in a fragment of the cranium, presenting their inner sides, the bone being here broken away. The lower parts of these teeth, in the neighbourhood of the crowns, as well as these, were hidden in the matrix when the specimen came into my hands. By carefully removing the matrix, the triturating surfaces of the three anterior cheek-teeth (the three premolars) were laid bare, and it became at once apparent that this fossil is a *Lagopsis*.

It was too late to have the teeth figured in the present memoir, so that I must content myself with their description. I give figures of them elsewhere †. The posterior of the three teeth (p. 1) exhibits the pattern, which is shown by the homonymous premolar of *Lagomys* and by the latter's two true molars. On the outer side of this tooth is a shallow and open groove, which, so far as can be made out under a strong lens, has no enamel border. From the middle of the inner margin a lozenge-shaped narrow enamel fold (*a* of my figures in Pl. 36) runs transversely across two-thirds of the breadth of the triturating surface towards the outer side; the posterior enamel border of this fold is raised into a strong crest, running parallel with the anterior enamel border of the tooth, both presenting a slight convexity turned anteriorly. The enamel fold is filled with cement in its outer narrower portion, its wider internal opening being devoid of this substance.

The pattern of the middle premolar, p. 2, proves that the isolated tooth from La Grive (Pl. 36. fig. 31) has been rightly determined as p. 2. As in the latter and in *Lagomys,* there is only a comparatively shallow internal enamel fold (*a*) present in the tooth from Œningen, the greater part of the triturating surface being occupied by the enamel crescent (*b*) before described in the tooth from La Grive. Outside the crescent (*b*) appears a small enamel ring filled, like the latter, with cement; this ring is doubtless the vanishing homologue of the outer enamel crescent (*c*) of *Titanomys* and *Prolagus,* described in the preceding pages and figured in Pl. 36. In the La Grive specimen (fig. 31) there is

---

* Catalogue of the Fossil Mammalia in the British Museum (Natural History). i. p. 256, No. 42815 (1885).
† Geol. Mag., dec. iv. vol. vi. p. 370, figs. 1 & 2 (1899).

a mere vestige of some such element in the same place, the tooth being presumably more worn than that in the Œningen specimen. As in the La Grive tooth, that from Œningen has both enamel margins of crescent (*b*) raised into triangular cusps, with the convexity turned inward.

The anterior premolar, p. 3, of the Œningen fossil is not dissimilar to the same tooth of *Prolagus œningensis* (Kön.). Whereas in recent *Lagomys* the triturating surface of p. 3 exhibits only one enamel fold—starting from about the middle of the anterior margin and running backward obliquely, *i. e.* postero-externally—the same tooth in *Lagopsis* shows two enamel folds, as in *Prolagus œningensis*, opening on the anterior margin, and thence running almost straight backward.

These differences from *Lagomys* strengthen, therefore, Schlosser's opinion, that the Miocene fossil is to be considered as a genus (*Lagopsis*) distinct from *Lagomys*. At the same time they present a further link in the gradual transformation of the tooth-pattern (*Titanomys*—*Prolagus*—*Lagopsis*—*Lagomys*—*Lepus*). which begins in the hindmost molar of Lagomyidæ and, gradually proceeding forward, stops at p. 1 in *Lagopsis* and *Lagomys*, and at p. 2 in *Lepus*.

Genus LEPUS s. l.

It would seem more rational to treat of the Miocene *Palæolagus* before *Lepus*, since there are strong reasons for the assumption that the former is the ancestor of the latter. On practical grounds, however, I think it more advisable to give the description of *Lepus* first, because we can fully understand the dentition of *Palæolagus* only after having dealt with the dentition of the young of the existing genus ; and because, on the other hand, the latter exhibits a further development of the modernization initiated by *Titanomys*.

Hensel, writing in 1856, stated that, contrary to the usual descriptions of authors, the upper molars of *Lepus* consist each of a single cylinder, which in the second, third, and fourth teeth is provided with a deep enamel fold, filled with cement and penetrating from the inner side [*]. When contending that all the previous writers on the subject had incorrectly interpreted the conformation of the leporine molar, Hensel could hardly have guessed that 13 years later he might have made an almost similar complaint. We continually meet with descriptions and figures of lagomorphous animals in which the upper molars are represented as formed by two cylinders closely united or soldered together, presenting three transverse enamel ridges !

As compared with the *Lagomyidæ*, by the presence of m. 3 in the maxillary, *Lepus* exhibits a more primitive condition. In the characters under consideration, however, *Lepus* is undoubtedly the extreme member of the series. While in *Lagomys* the posterior premolar (p. 1) has alone acquired the transverse fold of the true molars, in *Lepus* (Pl. **36**. fig. **33**) p. 2 has been transformed as well. P. 3 alone retains what we may fairly consider to be the ancestral enamel folds, as well as the ancestral internal notch. There is no anterior "wall"; wherefore the enamel folds open freely on the anterior side.

[*] Zeitschr. deutsch. geol. Ges. p. 681 (1856).

In a skull of *L. nigricollis* from Ceylon (B. M. Z.D. No. 81.4.29.7) (Pl. **36**. fig. 34) I find that the modernization has begun to invade p. 3 also; in the tooth of the right side the internal notch (*a*) has assumed the shape of a lengthened fold, stretching halfway across the crown and provided with plications as in the other molars.

M. 3 of *Lepus* is a small, vanishing cylindrical tooth; in rare cases, however, of *L. europæus* a transverse fold has been observed in this [*].

Now as to the condition of the teeth in the young of *Lepus*. Hilgendorf stated long ago [†] "that the upper check-teeth of young Hares are provided with a crescentic enamel tube, which forms a transition to the fossil *Myolagus*." This is perfectly true, but it is not all.

In the Rabbit *Oryctolagus cuniculus*, the two posterior upper deciduous teeth when worn, and the permanent molars when slightly abraded (Pl. **36**. fig. 5), exhibit a pattern identical to that presented by the two anterior true molars of *Palæolagus*, as figured in Pl. **36**. fig. 36, viz., an internal notch and a central crescentic enamel fold. Before attrition has set in, they exhibit besides a strong crescentic cusp (6), which delimits the outer side of the enamel fold (fig. 1). On the outer side of the cusp is seen a minute and shallow enamel fold, incompletely divided into an anterior and a posterior part by a ridge descending from the middle of the outer slope of the cusp (*c*, figs. 1, 5). In d. 2 the anterior horn of the larger crescentic enamel fold stretches further outward than in d. 1, and almost reaches the outer border of the tooth. When attrition is going on, the shallower outer fold may be seen for a short while on the triturating surface, under the form of one or two minute enamel islets, which are soon completely worn away. The deeper inner crescentic fold (*b*), apparently that mentioned by Hilgendorf, persists longer.

Here then we still meet with, in an ephemeral condition, the elements constituting the *Titanomys*-tooth: two enamel folds (*b* and *c*) separated by a strong cusp (6) and an internal notch (*a*). The deciduous teeth of *Lepus* s. l. are cast off without presenting any other change except that produced by further wear (fig. 26). In the permanent teeth (Pl. **36**. fig. 17) the internal notch begins to extend. That this growth takes place, in these initial stages at least, wholly in an inward direction—by a prolongation of the two internal cusps, which have gradually been transformed into transverse lobes [‡]— becomes evident when we compare these teeth before attrition and in a moderately worn condition. In the former stage the crescentic fold is separated from the internal

---

* Hilgendorf. in Monatsber. K. preuss. Akad. der Wiss. Berlin. 14 Dec. 1865, p. 673.   † *Ibid.*

‡ " 8 " and " 9 " in the figures of all the upper check-teeth on Pls. **36**, **37**, **39**. The scarcity of my material prevents me from entering into particulars with regard to the young stages of other recent Leporidæ. In a slightly abraded p. 2 of *Caprolagus hispidus* (Pl. **36**. fig. 27), *b* and *c* surround almost completely the well-developed cusp (6) and unite together to form a common outlet on the antero-external side of the tooth. The enamel exhibits numerous secondary plications characteristic of the teeth of this Hare. In the deciduous teeth of *Sylvilagus brasiliensis* (Pl. **36**. fig. 20), *a* and *b* are united and present the pattern of a branched fork, visible also in young stages of permanent teeth; in the latter (*b*), represented by the two branches of the fork, soon disappears from the triturating surface. In the true Hares, *Lepus* s. str. (Pl. **36**. figs. 22, 25, 28), the primitive pattern is more ephemeral still than in the Rabbit; the enamel crescent (*b*) is quite superficial. As is generally the case in disappearing structures, these vanishing elements present a considerable amount of variation in different specimens of the same species.

notch only by a longitudinal enamel ridge; in the latter it is still in its place, while the internal notch has grown into a transverse fold stretching across half the transverse diameter of the triturating surface *.

Upper Incisors of Leporidæ.

The upper incisors of several Leporidæ present some little-known peculiarities.

In his description of *Lepus nigricollis*, G. R. Waterhouse says:—" The upper incisor teeth have each two longitudinal grooves, placed very closely together, and not very distinct" †. About the same statement is made with regard to *Lepus yarkandensis*, Günth., by Büchner, who believes this to be a special character of the species:—" Sehr characteristisch für *Lepus yarkandensis* ist der Bau der oberen Nagezähne, durch welchen diese Art sich, wie es scheint, von allen Gattungsgenossen unterscheidet. Die Vorderfläche des oberen Backzahnes weist nämlich zwei flache, schwach markirte Rinnen auf; dieselben verlaufen dicht neben einander auf der inneren Hälfte der Vorderfläche" ‡.

I have before me the type-specimen of *L. yarkandensis*, Günth. (Br. Mus. Z. D. No. 75. 3.30.10); an examination of the outer surface of its upper incisors shows but one groove, as in other Leporidæ; the groove is filled with cement, but only incompletely, so that the outer and inner border of the zone of that substance is marked by two longitudinal striæ which somewhat simulate grooves. There is besides a median superficial depression of the cement layer, so that the appearance of three longitudinal grooves is produced. (In *Caprolagus hispidus* the median hollowing of the cement is more accentuated.)

In *L. nigricollis*, as a rule, the appearance of two grooves is produced by the same cause as in *L. yarkandensis*. Sometimes, however, there is in the former species a very shallow longitudinal groove in the enamel, to the outside of the principal groove filled with cement; the former is somewhat more distinct in the unique skull of a specimen from Ceylon in the Br. Mus. (Z. D. No. 81.4.29.7).

The fact of the presence of cement in the groove having been overlooked has given rise to another misunderstanding. Waterhouse says that in *Lepus ruficaudatus* the

---

* According to Père Heude, the anterior upper premolar, p. 3, of *Lepus* is composed of p. 3 and a more anterior premolar, which latter is said to be represented by the median of the three anterior lobes (" 6 " of my figures) of p. 3. (*op. cit.* pp. 63, 64. pl. xiii. figs. 4, 5, 7. 1898). As I believe to have satisfactorily demonstrated—although not, perhaps, to the Rév. Père's satisfaction—that this " 6 " of p. 3 is the homologue of " 6 " in the posterior premolars and true molars of all Lagomorpha, I think we can, for this reason alone, dismiss the fusion theory, since each of these posterior teeth would have to be considered also as a compound of two. (Similar remarks apply to p. 2 of the lower jaw of *Lepus*, which, according to Père Heude, is = p. 2 + p. 3.) I may add here that I have never observed in the upper molars or premolars of *Lepus* a longitudinal enamel ridge closing the opening of the internal enamel-inflection (*a* of my figures), as figured and described by Père Heude (" fissure qui se ferme avec une lamelle d'émail chez l'adulte," *op. cit.* p. 65, pl. xiii. fig. 4), and would gladly learn in which species this occurs.

† G. R. Waterhouse, ' A Natural History of the Mammalia,' ii. p. 73 (1848).

‡ Eug. Büchner, ' Wiss. Resultate der von N. M. Przewalski nach Central-Asien unternommenen Reisen,' i. 5. p. 193 (1894).

superior incisor " has the external groove less deep, and placed nearer to the inner edge of the tooth," than in the Common Hare * ; and W. T. Blanford states of *Lepus dayanus*, Blf., that "the upper incisors appear very indistinctly grooved " †. The species mentioned are precisely among those in which the groove of the upper incisors is very deep ; but they present the appearance of being shallow, owing to the cement which incompletely fills them. In fact, the cement appears in all species in which the groove penetrates further backward than in the commonly accessible species (*L. europæus*, *Oryctolagus cuniculus*), and it is in that case very often associated with other complications which we have now to consider.

Hodgson gives as one of the distinctive characters of *Caprolagus hispidus* the following :—" the groove in front of the upper incisors is continued to their cutting-edge so as to notch it " ‡. Strictly speaking, the cutting-edge of the upper leporine incisors is always notched—even in *Lepus europæus* ; only, in *C. hispidus* (text-fig. VIII), the groove, filled with cement, is much broader and penetrates further backward, so that the natural section presented when the incisor is viewed from its lower side (same fig.) shows the groove under the form of a very elongated triangle, with the apex at its posterior end. A more complicated form has been noticed by Hilgendorf, as stated in the following brief sentence :—"Die oberen Schneidezähne von *Lepus callotis* aus Mexico und *Lepus nigricollis* aus Indien sind gabelig schmelzfaltig (dentes complicati) ; die entsprechenden Zähne der afrikanischen Hasen bilden durch eine einfachere Einbuchtung des Schmelzes einen Uebergang von jenen zu den anderen Hasenarten " §. In a later note by the same writer further particulars are given ‖. In the text-figures I–XXIV are shown, enlarged (about 4×1), the principal modifications of the enamel-folding of upper leporine incisors viewed from below and with the anterior border directed downward. Some slight differences between the few descriptions given by Hilgendorf and my figures of the supposed same species are apparently due to different causes : in the first place, because Hilgendorf describes tooth-sections. Moreover, specimens of the same species may vary slightly (*cf.* figs. XVI & XVII), owing partly to individual variation. But the shape of the enamel-fold varies equally at different stages in the age of the animal ; species whose incisors show the most complicated pattern in the adult have as yet no trace of this in very young animals ; and, *vice versâ*, in very old specimens complication tends to disappear again. As shown by several of the text-figures, slight variations between the right and left incisor of the same individual also occur. These circumstances will, of course, have to be taken into account for systematic purposes.

The most complicated folding in Hilgendorf's material was presented by a *L. callotis*, Wagn. (=*L. mexicanus*, Lichtenst.), from Mexico ¶, in the shape of a **T**, whose transverse

---

* *Op. cit.* p. 77.—R. Swinhoe (Proc. Zool. Soc. Lond. 1870, p. 234) makes a similar remark with regard to *L. hainanus*.

† W. T. Blanford, "On New Mammals from Sind," P. Z. S. London (1874). p. 663.

‡ Journ. As. Soc. Bengal, xvi. 1, p. 576 (1846).

§ Sitzungsber. Berl. Ak. Wiss., Sitzg. 14 Dec. 1865 (1866).

‖ Sitzungsber. Ges. naturf. Freunde Berlin, Sitzg. 15 Jan. 1884, pp. 18–21.　　　　　¶ *Op. cit.* pp. 18, 19.

part, turning backward, runs approximately parallel with the anterior border of the tooth, and is slightly folded from behind, so that it may be compared with an outspread fork. Figs. XVI and XVII, representing the left incisors of two specimens from Mexico in the Nat. Hist. Museum, labelled *Lepus callotis*, show this same form, with a slight complication of the transverse part in one of them (XVII). *L. melanotis*, Mearns (fig. XV), from Clapham, New Mexico, belonging to the same group (*Macrotolagus*), exhibits in the right incisor the **T** pattern in a much reduced form, and in the left a condition approximating to that of the African *L. saxatilis*, of which more hereafter.

The nearest approach to *L. callotis* is seen, according to Hilgendorf, in *L. dayanus*, Blf., to which species he refers also the *L. nigricollis* of the first note. I have figured (fig. XVIII) the right incisor of the co-type of *L. dayanus*, from Sukkur, Sind (Br. Mus. Z. D. No. 90.6.9.2), which corresponds almost exactly to Hilgendorf's description. A nearly similar form I find to be exhibited by *L. hainanus*, from Hainan (fig. XIX); the folding, however, is considerably shorter, and the opening broader. In *L. nigricollis* from Ceylon (fig. XXI) the branches of the fork are more elongate, and the anterior opening is considerably more constricted, than in *L. hainanus*.

L. peguensis, Blyth, from Pegu (fig. XX), shows a further complication, already foreshadowed by one of the *callotis* specimens (fig. XVII), there being three branches of the fork. Not much different is the left incisor—the right one is damaged—of a *L. nigricollis* from the Nilghiris (fig. XXII), and both incisors of *L. ruficaudatus* (*L. kurgosa*, Gray) from the Punjab (fig. XXIII). The maximum of complication known to me is exhibited by a *L. ruficaudatus* from Rajputana (fig. XXIV), where the left incisor exhibits a four-branched fork, the right being a slight modification of the same pattern.

Following the description of the incisors of *L. dayanus*, Hilgendorf gives that of an undetermined skull brought home from Africa by the Von der Decken Expedition. In this the **T** with a narrow opening is still strongly marked, but the median moiety of the transverse part is reduced. The whole of the enamel-fold occupies less space than in *L. dayanus*, not being so much extended either backward or laterally [*]. This description applies fairly well to my fig. XIV, *L. Victoriae*, Thos., from Nassa, Victoria Nyanza, except that the opening of the fold is not narrowed.

Figs. IX, X, and XII represent *L. saxatilis*, F. Cuv., from Pirie Bush, King William's Town (Cape), Transvaal, and " Cape of Good Hope " respectively, in none of which is there a bifurcation at the posterior end; the folding penetrates far backward and the opening is wide, as described by Hilgendorf [†] in *L. saxatilis*. Fig. XI, " *Lepus* sp.", from Sena, Zambesi, is of the same pattern; and so is *L. Whytei*, Thos., type-specimen, from Pacombi River, Nyasa (fig. XIII); in the latter, however, the fold penetrates further back than in figs. IX–XII, and the opening is comparatively more restricted. To this form seems to approach Hilgendorf's specimen of " *Lepus capensis*,"

* Op. cit. p. 20. † Op. cit. p. 21.

65*

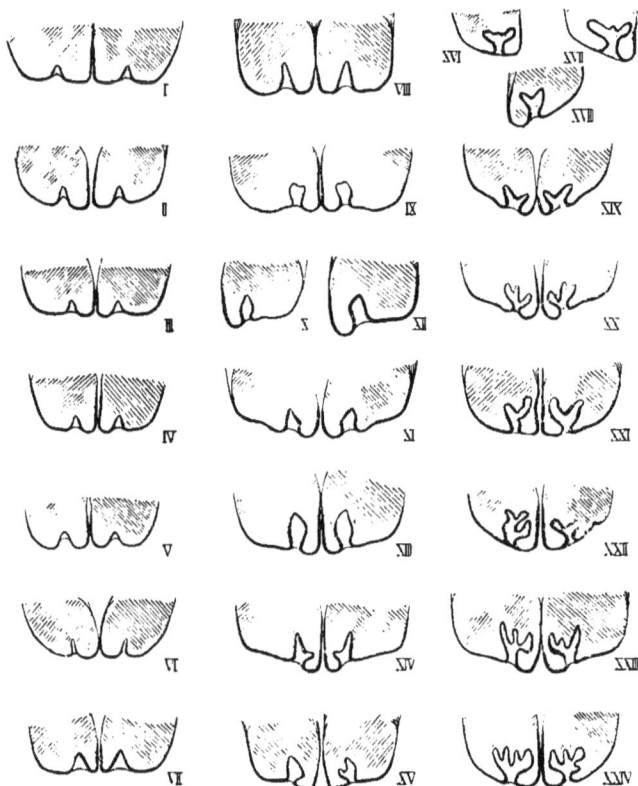

Anterior end of upper Leporine incisors, from below.   Enlarged.

No.			Brit. Mus. Z. D.	
I.	Lepus variabilis (altaicus).	Russia.	535 a.	40.5.13.4.
II.	L. variabilis. ♀.	Altyre, Morayshire.	„	—— 94.2.15.2.
III.	L. sinaiticus.	Midian, N.W. Arabia.	„	—— 78.8.9.1.
IV.	L. Judææ, Gray, ♀, type.	Palestine.	„	—— 64.8.17.5.
V.	L. sinensis, Gray, type.	China.	„	—— 38.10.20.23.
VI.	L. cumanicus, Thos., type.	Venezuela.	„	—— 94.9.25.18.
VII.	" L. yarkandensis ? "	Koko Nor.	„	—— 94.2.2.12.
VIII.	Caprolagus hispidus, Pears.	(Ind. Mus. Coll.—B. H. Hodgson.)	„	—— 79.11.21.204.
IX.	Lepus saxatilis.	Pirie Bush, King William's Town (Cape).	„	—— 98.10.8.1.
X.	L. saxatilis, ♂.	Transvaal.	„	—— 93.11.26.2.
XI.	Lepus sp.	Sena, Zambesi.	„	—— 83.2.6.3.
XII.	L. saxatilis	C. G. Cope.	„	525 a. 42.12.6.5.
XIII.	L. Whytei, Thos., ♀, type.	Pacombi River, Nyasa.	„	—— 94.1.25.14.
XIV.	L. Victoriæ, Thos.	Nassa, Victoria Nyanza.	„	—— 95.3.7.2.
XV.	L. (Macrotolagus) melanotis, Mearns, ♂.	Clapham, New Mexico.	„	—— 94.5.9.29.
XVI.	L. („) callotis.	Mexico.	„	—— 58.9.22.2.
XVII.	L. („) callotis.	Mexico.	„	—— 53.8.29.37.
XVIII.	Lepus dayanus, co-type.	Sukkur, Sind.	„	—— 90.4.9.2.
XIX.	L. hainanus.	Hainan	„	—— 70.7.18.18.
XX.	L. peguensis, Blf., ♀.	Pegu.	„	—— 91.5.12.1.
XXI.	L. nigricollis, F. Cuv.	Ceylon.	„	—— 81.4.20.7.
XXII.	L. nigricollis.	Kotagiri, Nilghiris.	„	—— 91.10.7.154.
XXIII.	L. ruficaudatus (kuryosa, Gray).	Punjab.	„	1176 a. ——
XXIV.	L. ruficaudatus.	Rajputana.	„	—— 91.10.7.151.

from Mozambique *, collected by Peters, which, however, is certainly not a *Lepus capensis*. The latter differs scarcely from *L. europœus*, Pall., s. l. (including *L. occidentalis*, de Wint.), by its minute enamel-folding, not filled with cement.

The forms which remain to be described (figs. I-VII) are all approximately of the same type, viz. a triangular fold with the apex turned backward; the fold in none of them stretching so far back as in *Coprolagus hispidus* (fig. VIII), mentioned above. The pattern of the latter is approached somewhat by that of fig. VII, from a specimen labelled " *Lepus yarkandensis?*," from Koko Nor (Br. Mus. Z. D. No. 94.2.2.12), exhibiting an enamel-fold with thick borders, but shorter than in *C. hispidus*, and with a much wider opening. It is decidedly not *L. yarkandensis*, Günth. The type of the latter, which is not figured, approaches in the form of the folding *L. sinensis*, Gray, the type of which (Br. Mus. Z. D. No. 38.10.29.23) is represented in fig. V. Both are imperfectly filled with cement, in *L. sinensis* still less so than in *L. yarkandensis*. The latter differs also from the former by the opening and the whole fold being narrower.

L. tibetanus, Waterh., has no trace of cement ; in the shape of its fold it is intermediate between the former two ; the opening is slightly broader than in *L. yarkandensis*.

The conformation of the type of Gray's " *L. Judeæ* " (fig. IV), from Palestine, and of " *L. sinaiticus* " (fig. III), from Midian, N.W. Arabia, almost identical in both, is shown by the figures.

L. timidus, Linn. (*L. variabilis*, Pall.) (figs. I & II) hardly differs, but still the two figures of this species show that there are slight differences between a specimen from Russia (fig. I) and one from Scotland (fig. II). In this species I have always found the enamel-fold with a filling of cement, though very often incomplete. In *L. europœus*, Pall., I have never met with a trace of cement. This difference would seem to be a good character for distinguishing isolated fossil incisors of the two species ; but it is probable that much-weathered specimens of *L. timidus* may have lost their cement.

Lepus cumanicus, Thos., from Venezuela (Br. Mus. Z. D. No. 94.9.25.18), the type of which is represented in fig. VI, stands somewhat apart by its very narrow and comparatively elongate enamel fold.

Hilgendorf holds these complications of the enamel in the upper incisors to be a specialization, the only reason given being that in the fossil *Prolagus* (*Myolagus*) nothing of the kind is seen. " Phylogenetisch betrachtet, ist die bedeutende Schmelzentwicklung des *Lepus mexicanus* gleichfalls ein Extrem ; denn die Einbiegung der Schmelzplatte an der Vorderfläche tritt bei den fossilen Leporiden-Gattungen (*Myolagus*) als eine seichte Einknickung auf, deren Seitentheile fast die ganze Vorderfläche einnehmen " †. This argument would be of some weight if *Prolagus* could be considered ancestral to *Lepus*; but this is certainly not the case, although the molars of the former are of a more primitive type than those of the latter. As insisted upon in the present memoir, the Lagomyidæ, of which *Prolagus* is a member, run parallel with the

---

* *Op. cit.* p. 21.   † *Op. cit.* p. 20.

Leporidæ from the Lower Miocene (or it may be from the Oligocene) to the present time.

The incisors provided with enamel-folds point back towards cuspidate incisors, for the enamel-folds of lophodont and laminated teeth are obviously the derivates and homologues of the "valleys" separating the cusps or tubercles. Now it is very suggestive that we meet with cuspidate incisors in *Plesiadapis*, a genus from the lowest Eocene of Rheims, classed among the Lemuroidea by Lemoine and other writers, considered by Schlosser and me to be a very primitive Rodent. In the jaws of *Plesiadapis* the teeth are greatly reduced in number. In the lower jaw we have only one powerful elongated incisor, directed obliquely forward and upward, and separated from the five cheek-teeth—the premolars being already reduced to two—by a considerable diastema. On its posterior face the lower incisor has a cingulum supporting a small cusp. The upper incisors, too, are separated by a long interval from the five cheek-teeth, and appear to have been three in number (Lemoine considers the very small outer one to be the canine). The two outer pairs are very small and unicuspidate; the inner pair robust, generally tricuspidate, there being an anterior pair of cusps, and backwardly an additional cusp, which starts from a kind of cingulum *.

If we imagine the cusps of these upper incisors of *Plesiadapis* to have become lengthened in accordance with a general change of the more brachyodont incisors into a hypselodont one, and their interstices filled with cement, so that by trituration a level surface can be produced, the result would be a pattern somewhat similar to that of several of the figured Leporidæ. The posterior cusp of *Plesiadapis*, projecting from behind into the cavity †, would produce a posterior ramification like that of the Leporidæ.

The test will lie in the search for Tertiary Leporidæ exhibiting an intermediate stage between the condition of the upper incisors of *Plesiadapis* and that of recent Leporidæ. An examination of the incisors of *Palæolagus* might decide the question.

Genus PALÆOLAGUS.

Palæolagus, from the Tertiary of North America, is represented by Leidy ‡ and by Cope § as showing in the teeth only one character distinctive from the genus *Lepus*, viz. the more simple conformation of the anterior inferior premolar of the extinct genus, and of this character more hereafter. When, however, we go over the descriptions, accompanied by numerous figures, and an examination of originals, several of which are in the British Museum, we cannot but be struck at once by some very essential differences in the triturating surfaces of the two genera. When do we ever meet with molars in any species of *Lepus* showing the complete absence of all traces of enamel, with the exception of part of the marginal border? This is the case in old

---

* Lemoine, in Bull. Soc. Géol. France, xix. 1. p. 278, pl. x. fig. 50, *a*, *b*, *c* (1891).

† Lemoine, *l. c.* pl. x. fig. 50, *b*, *c*.

‡ Proc. Acad. Philadelphia, p. 89 (1856); id. 'Extinct Mammalia of Dakota and Nebraska,' p. 332, pl. xxvi. figs. 14-20 (1869).

§ 'The Vertebrata of the Tertiary Formations of the West,' i. p. 870, pls. lxvi., lxvii. (1883).

specimens of *Palæolagus*. The pattern of less worn teeth, too, is rather different from what occurs in *Lepus*. In none of the numerous triturating surfaces of *Palæolagus*-teeth figured do we meet with a transverse fold penetrating so far outward as in the four intermediate teeth of *Lepus*, and in the true molars and posterior premolar of *Lagomys*. This is confirmed by Cope's description:—" The inner side of the four intermediate molars is deeply grooved *for a short distance* " (italics mine; *cf*. Cope's figures), " which gives a fissure-like notch on attrition. This disappears after use, as does also a less profound crescentic fossa in the middle of the crown, whose concavity is directed outward " *.

This statement, in my opinion, does not fully describe the pattern in young specimens, which seems to be very ephemeral in *Palæolagus*. In a fragment of the right upper jaw of *P. Haydeni* in the Brit. Mus. (5727), of which I give an enlarged figure (Pl. **36**. fig. 36), the alveolus of the second premolar (p. 2) is shown, and the three teeth p. 1, m. 1, m. 2 are seen in place. The empty alveolus of the premolar suggests that in its contour this tooth very much approached the corresponding tooth of *Prolagus œningensis* (Pl. 36. fig. 21), and to judge from what we find in the following tooth (p. 1) there is a strong assumption that the pattern of p. 2 of *Palæolagus* also resembled that of *Prolagus œningensis*. P. 1 of *Palæolagus* exhibits the internal notch (*a*) with which we are acquainted in *Titanomys* and in the deciduous teeth of *Prolagus*, *Lagomys*, and *Lepus*, and which moreover persists as such in the premolars of *Prolagus*, in the second premolar of *Lagomys*, and in the anterior premolar of *Lepus*. In the premolar of *Palæolagus* we find, on proceeding inward, a crescentic central enamel islet in the centre of the crown, known already from the descriptions and figures of Leidy and Cope. It is, too, an old acquaintance of ours; for to all appearance it is the homologue of the large internal enamel-fold (*b*) of *Titanomys*, whose further history we have followed up in the other genera. But this is not all. From the antero-external corner of p. 1 of *Palæolagus* starts an enamel-fold in a postero-internal direction, terminating near the outer end of the crescentic fold's posterior horn. No mention is made of this outer fold in Leidy's and Cope's descriptions; it is, however, visible in one p. 2 of Cope's figures (pl. lxvii. fig. 16 *a*); but I have not seen it delineated for the same tooth together with the crescent fold, as in the figure which I publish. The outer fold just described is undoubtedly the homologue of the outer enamel-fold (*c*) of *Titanomys*, and I do not doubt that still younger stages of *Palæolagus*—which have been figured by Cope, but in an unsatisfactory manner—will show a greater development of both the enamel-folds, and therewith a stronger resemblance to the pattern of the *Titanomys*-teeth and the posterior premolars of *Prolagus*.

The true molars of *Palæolagus* in the specimen figured exhibit only the crescentic central islet (*b*) and the internal notch. As stated by Cope in the passage quoted above, and as shown likewise by the illustrations of both the American writers, the internal notch and the crescentic islet are worn away by attrition, without any other change taking place. In this consists the great difference between the American fossil and all the forms

* *Op. cit.* p. 876.

previously described in this paper. While in all the upper grinding-teeth of *Titanomys* the initial condition, two crescent folds and an internal notch, is retained throughout life, and this is more or less so in the premolars of *Prolagus*, in the molars of the latter the crescentic folds are worn away and the internal notch is enlarged to a transverse fold, s in the molars and p. 1 of *Lagopsis* and *Lagomys*, and in the molars and posterior premolars of *Lepus* s. l. Milk-teeth and very young permanent molars of *Lepus* show, with slight variations, the pattern before described as characteristic of moderately-worn teeth of *Palæolagus*. No modernization takes place in the latter ; the only change we perceive, by the further progress of wear, is the complete obliteration of the crescentic folds and of the notch on the inner side. In *Lepus*, the large crescentic fold of the deciduous teeth, and a small islet external to it—present in some of the species, and representing the *external* crescentic fold of Lagomyidæ—disappear at a very early stage of the two posterior premolars and of the two anterior true molars, and are replaced in the permanent teeth by the transverse fold already described.

The permanent teeth of *Palæolagus*, therefore, can only be compared with the deciduous teeth of *Lepus*; like these (Pl. 36. fig. 26), they exhaust their primitive pattern, without evolving a secondary one [*].

Palæolagus cannot find a place in our phylogenetic series (*Titanomys*—*Prolagus*—*Lagopsis*—*Lagomys*). With regard to the condition of their upper cheek-teeth, the species of *Palæolagus* in which these teeth are known would follow after *Titanomys*. But they are certainly not the forerunners of *Prolagus*, except in the form of the true molars ; while *Prolagus* is more conservative than *Palæolagus* in the conformation of its two posterior premolars. On the other hand, *Palæolagus* is certainly the forerunner of *Lepus*, and presumably its ancestor ; and this cannot be said of the *Lagomyidæ*, in all of which the upper m. 3 has been lost.

To resume.—We have in the preceding pages followed the transformation in the pattern of the upper cheek-teeth on three lines :—(1) From genus to genus ; (2) from behind forward in the dental series ; (3) from young to old.

(1). *From genus to genus*, we might almost say from species to species, the series is as follows:—*Pelycodoid* type (*Pelycodus, Plesiadapis*)—*Titanomys risenoviensis*—*T. Fontannesi*—*Palæolagus*—*Prolagus œningensis*—*P. sardus*—*Lagopsis*—*Lagomys*—*Lepus.*

Pelycodus and *Plesiadapis* are genera of the Lower Eocene.

Titanomys appears in the Lower Miocene, and vanishes in the Middle Miocene.

Prolagus appears in the Middle Miocene and lingers on, protected by an insular habitat, until the Neolithic period.

Lagopsis is at present known only from the Middle Miocene.

Lagomys makes its appearance in the Pleistocene and survives to the present day.

Lepus, preceded by the Oligocene and Miocene *Palæolagus*, appears with many of its present generic characters in the Lower Pliocene, and survives to the present day.

---

[*] The remarkable Hare from Sumatra, *Nesolagus Netscheri*, approaches *Palæolagus* more than other recent Leporidæ, inasmuch as, by the feeble development of the transverse enamel-fold (Pl. 37. fig. 17), it represents a first stage in the evolution of the secondary pattern. The same form exhibits other primitive features, to be described later on.

(2) *From behind forward in the dental series.*—The true molars are the first to be transformed, and successively one after the other of the premolars, the anterior premolar (p. 3) being the most conservative.

(3) *From young to old.*—The cheek-teeth of the genera under consideration exhibit, in the first developed parts of their shaft, more or less evident traces of the ancestral pattern; *mostly so* the deciduous teeth, which are cast off when the primitive pattern has almost vanished, and without showing a beginning of transformation; *least so* the true molars, which in the first stages observable of the calcified tooth, and before trituration has set in, show the primitive pattern already reduced and the secondary in process of evolution.

LOWER MOLARS OF LAGOMORPHOUS RODENTIA.

To state it in a general way, the lower molars of the Lagomorpha present the same characters as their upper antagonists : viz. anteriorly in the series we meet with complication, posteriorly with a simple transverse pattern. On closer examination, however, it may be seen that in the mandibular teeth the process which we have followed through its various stages in the upper set is accelerated. Although it must be taken into account that we have one premolar less below than above, none the less—leaving for the present out of consideration the reduction which takes place at the posterior end (m. 3)—there is in the adult mandible only one tooth, the anterior, which differs materially from the others, by being more complicated. In *Titanomys*, the oldest member of the group, this tooth (p. 2) as generally described and figured, presents a more simple structure than in later genera, and even than do the other teeth of *Titanomys*, by being composed of only one column, divided into two lobes by an inner and an outer enamel-inflection : whereas in the teeth situated posteriorly there are two columns, the division between them being complete : they are held together by cement.

We meet here with a phenomenon which is pretty general among Rodents, whether the number of their cheek-teeth be three, four, five, or six. To state it more fully :—

1. The mandibular cheek-teeth precede those of the maxilla in the reduction of their number ; we have instances of $\frac{6}{5}$, of $\frac{5}{4}$, and of $\frac{4}{3}$ cheek-teeth, but never of $\frac{5}{6}$, or $\frac{4}{5}$, or $\frac{3}{4}$.

2. Very frequently the anterior tooth in the lower series, whether it be p. 2, or p. 1, or m. 1, is more complicated than those behind ; which circumstance suggests that the complication has some connection with the anterior position of the tooth in question.

3. When the anterior lower tooth is nearly or actually equal in pattern to those behind, this is generally so in older forms. Thus we find that in Winge's Anomaluridæ—including mostly Tertiary genera—provided with four lower teeth, the anterior one (p. 1) is equal or subequal in size and pattern with the others, and sometimes even of smaller size. Again, in Muridæ, with three inferior cheek-teeth, the geologically older forms have the anterior one (m. 1) equal or subequal in size with the two following, whereas the complication of the first molar appears only in more recent forms. The same is true with regard to the lagomorphous Rodentia, where the anterior tooth is p. 2, and in the oldest known genera (*Titanomys, Palæolagus*) of a rather simple pattern.

The explanation which I suggest for these curious occurrences is as follows :—When an anterior tooth drops out from the mandible—generally through an apparent interference of the incisor with its pulp—some compensation for its loss is necessary, as the corresponding maxillary tooth is generally still in its place; this compensation is brought about by a complication on the anterior side of the tooth which has become the first in the series by the loss of the originally anterior one. Those genera which are nearer in date to the epoch when the anterior tooth was lost will still present a less complicated form of that which has succeeded to this position, while in the later genera the foremost tooth will have acquired the complication. When p. 2 is dropped, p. 1 will become the foremost tooth, and the same cycle will recommence, and so on.

I next proceed to a closer examination of the lower cheek-teeth, starting from those of *Titanomys*. A superficial comparison of the anterior tooth, $\overline{p.2}$, of this genus, with that of the other Lagomyidæ, shows that in the former it is more simple than in the latter, and presents an approximately tetragonal outline at its triturating surface; in *Prolagus*, *Lagopsis*, and *Lagomys* this is triangular (apex in front). Thus it is that we find the tooth generally described; but on closer examination the matter is somewhat more complex. I have figured five specimens of p. 2 of *Titanomys Fontannesi*, from La Grive-Saint-Alban, in different stages of wear; four are isolated teeth (Pl. **37**. figs. 1–4); the fifth is in its place in a left ramus, presenting the complete series of two premolars and three molars (Pl. **37**. fig. 7). Of *T. eisenoriensis* I have one specimen, in a fragment of the right ramus, containing the two premolars (Pl. **37**. fig. 25). This species is from the Allier (Bravard Collection, Br. Mus. Geol. Dep. No. 31095). The first stage in *T. Fontannesi* (fig. 1) represents a tooth which has not yet come into wear. In the main it is composed of two lobes; the anterior is subconical, the posterior is much more extended transversely, and composed of a tapering outer and a thicker, rounded inner cusp; moreover, on the middle of its posterior surface appears a small cusp (*t*); the anterior surface of this lobe is wrinkled. Even in this early stage the separation of the two lobes is incomplete; a ridge, running almost longitudinally backward, from the middle of the posterior side of the anterior lobe towards the posterior, shows that trituration would very soon have connected the two by a narrow isthmus of dentine, thus separating from each other an outer and an inner enamel-inflection. This we see, in fact, brought about in the second stage (fig. 2). Towards the middle of the anterior margin of the anterior lobe, a feeble cusp is visible in the first stage (1, fig. 1); the same is more distinct in the second stage (1, fig. 2), where it is nearer to the inner side. This cusp, to all appearance, is Winge's 1, Osborn's *paraconid*. Whether it contains potentially some other element I must leave undecided; as a matter of fact, in the two teeth described, it does not occupy exactly the same position; and in *T. eisenoriensis* (1, fig. 25) it is more approximated to the outer side. What is called the paraconid is, however, somewhat inconstant in its position [*]. In p. 2 of *T. eise-*

---

[*] See, *e.g.*, the text-figures in W. D. Matthew, " A Revision of the Puerco Fauna," Bull. Am. Mus. Nat. Hist. ix. (1897).

noriensis (fig. 25) it is evident as a small vertical pillar, lying far below the triturating surface of the moderately worn tooth.

To return to the second stage in *Titanomys Fontannesi*. The inner of the two principal enamel-inflections resembles somewhat in outline its homologue in *Lagopsis verus* (Pl. 37. fig. 26, p. 2). It is seen to be composed of two parts: a posterior, which communicates by a narrowed opening with the internal margin of the tooth, and thence runs straight towards the middle of the tooth, and an anterior circular one; the two communicating with each other by a narrow channel. The terminal cusp (t) is situated much nearer the inner side than in the first stage. I have dealt with this terminal cusp of the lagomorphous Rodentia on a former occasion, and homologized it with Osborn's *hypoconulid* [*]; a view from which I see no reason to depart. In the third stage (fig. 3) this hypoconulid is still apparent; but the "paraconid" has disappeared, and so has the circular part of the inner enamel-inflection. The transverse posterior part of the latter is on its way to be shut off from the inner margin, and to assume the form of a circular enamel islet. "t" is visible on the posterior internal edge of the tooth. In the fourth stage (p. 2 of fig. 7), the circular enamel islet is quite separated from the inner margin, and has become confluent with the outer enamel-inflection, so that the triturating surface of the tooth presents—if we except a small enamel fold limiting anteriorly the still extant t—only one enamel-inflection, penetrating from the middle of the outer margin and approaching the inner. In the fifth stage (fig. 4) we find only the latter inflection, t also having disappeared. This tooth in its general outline again approaches the first stage.

No lower deciduous teeth of *Titanomys* are at my disposal. Filhol has figured d, and d, of *T. eisenoriensis* from Saint-Gérand-le-Puy (Allier); from this figure nothing more can be made out than that in d. 2 the anterior part seems to be more produced anteriorly than in p. 2. No description is given of the triturating surface [†].

The anterior lower premolar of *T. eisenoriensis* is distinguished from the same tooth in *T. Fontannesi* by the persistence of the enamel-inflection of the inner side in the adult (Pl. 37. fig. 25); in the immature specimen figured by Gervais, and originally described as a separate species, *T. trilobus*, the two enamel-folds are confluent in the middle of the triturating surface, thus completely separating an anterior and a posterior lobe [‡]. The terminal cusp (t) present in the specimen figured (Pl. 37. fig. 25) must certainly be expected to be visible likewise in younger specimens; Gervais makes no mention of it in this tooth; in the profile view of the tooth, however [§], there are two vertical grooves on the inner side. A small anterior pillar ("paraconid") on the anterior side (1), below the triturating surface, has already been mentioned as present in the British Museum specimen.

---

* Proc. Zool. Soc. London, 1893, p. 203.

† H. Filhol, "Études des Mammif. foss. de Saint-Gérand-le-Puy, Allier," Ann. Sc. Géol. x. p. 29, pl. iii. fig. 3 (1879).

‡ Zool. et Pal. Fr. p. 51: "les deux lobes de la première molaire n'y sont point encore réunis l'un à l'autre par un petit isthme d'ivoire"; pl. 46, fig. 1 (1859).

§ Op. cit. pl. 46. fig. 1 c.

We have to follow up this same tooth, $\overline{p}.2$, in the other genera of Lagomyidae. In *Lagopsis cerus* (Pl. 37, fig. 26), from the Middle Miocene of La Grive-Saint-Alban, the posterior transverse lobe of p. 2 is undivided, with no trace of *l*. The next anterior lobe is separated from the former by a T-shaped enamel-inflection on the inner side—which has already been mentioned as approaching in form its homologue in *Titanomys Fontannesi* (fig. 2)—and by an outer one. We have, therefore, here the two enamel-inflections of *T. visenoriensis* and of the young of *T. Fontannesi*. However, in *Lagopsis* the lobe is more distinctly divided than even in fig. 2 (*T. Fontannesi*), into an outer and an inner cusp; for in the former the T-shaped inflection extends more anteriorly, and the lobe is delimited in front by two smaller enamel-folds. These latter delimit on their anterior side two further cusps, an outer and an inner; the latter corresponds to 1 (paraconid), as seen by comparison with fig. 2; the former may correspond to the pillar which in *T. visenoriensis* (fig. 25) is nearer the outer than the inner side. In any case, in *Lagopsis* the anterior part of p. 2 is much more developed than in *Titanomys*; for we have, in the former, two comparatively stout cusps against one feeble cusp in each of the two species of the latter. Besides, there is in *Lagopsis* a small odd cuspidule, situated in front of the anterior pair, and in the middle line of the tooth, to which it gives a triangular form.

The principal difference in *Lagomys*, to which *Lagopsis* is nearly related, consists in the fact that the characteristic T-shaped inflection of the *Lagopsis* p. 2 is either absent or replaced by a slight indentation of enamel. The latter is the case, *e.g.*, in *Lagomys rutilus* *, the former in *L. alpinus* and *L. nepalensis* †. Moreover, the odd anterior cuspidule has vanished in *Lagomys*.

In *Prolagus* also the anterior part of d. 2 is much more complicated than in p. 2 of *Titanomys*. Fig. 5, Pl. 37, shows this tooth of *Prolagus sardus*, var. *corsicanus*, from the ossiferous breccia of Toga, near Bastia (Br. Mus. Geol. Dep. No. M3486); fig. 6, the same tooth of the Miocene *Prolagus œningensis* from La Grive-Saint-Alban; both from the left side. I have still younger stages than those figured of this deciduous molar, showing the posterior lobe completely separated from the middle one. The anterior lobe of d. 2 of *P. œningensis* (fig. 6) is tripartite, as in *Lagopsis*, but the odd anterior cuspidule is less distinctly divided from the inner than in the latter genus. In the tooth of *P. œningensis* the whole tripartite lobe is connected only by cement with the rest of the tooth; in younger stages it is still more divided into a smaller external cusp—which is isolated, also, in the d. 2 of *P. sardus* figured (fig. 5)—and a larger internal one comprising both the " paraconid," 1, and the odd anterior cuspidule. The isolated small external cusp of *P. sardus* is situated far below the triturating surface; the inner larger one, showing no separated odd cuspidule, is connected on its inner side with the rest of the tooth, as happens likewise, though very rarely, in the corresponding permanent tooth, p. 2, of the same genus. In still more advanced stages

---

* For a figured specimen of this tooth see E. Schaff, " Ueber *Lagomys rutilus*, Sevortzoff," Sep.-Abdr. aus Zool. Jahrb. ii. p. 69, fig. 5 b.

† R. Hensel, " Beiträge z. Kenntn. fossiler Säugethiere," Zeitschr. deutsch. geol. Ges. viii. pl. xxvi. figs. 2 & 6 (1856).

of wear of the deciduous tooth of *Prolagus*, the whole of the anterior tripartite lobe appears invariably connected with the posterior part of the tooth by a dentinal isthmus, thus giving the whole tooth some resemblance to m. 1 inf. of a vole; and it has, in fact, been mistaken for a molar of *Microtus*.

A characteristic feature of the anterior lower premolar, $\overline{p}.\ 2$, of *Prolagus*, is an odd isolated cusp or pillar, connected only by cement with the rest of the tooth, and situated on its anterior side, thus giving to the whole tooth a triangular outline, as in *Lagopsis*. In *Prolagus œningensis* (Pl. 37. fig. 9) this cusp is situated near or close to the middle line; in *P. sardus* *, of which I have examined hundreds of specimens, its position is nearer the inner side. As before mentioned, in very rare cases of *P. sardus*, this usually isolated cusp is united with the tooth near the inner side, as in d. 2 of fig. 5. In other cases of *P. œningensis* (fig. 12, Pl. 37.) and *P. sardus*, it may be united with the tooth near its outer side. This latter fusion I found to have taken place in 19 specimens of p. 2 out of 575 examined, from the ossiferous breccia of Monte San Giovanni (Sardinia) (*P. sardus*), and in two cases out of 84 examined from Toga, near Bastia (*P. sardus*, var. *corsicanus*). The cusp was united with the tooth near the inner side in two of the 575 examples from Monte San Giovanni. Cusp "*t*" I have met with only in p. 2 of *Prolagus elsanus* (page 460).

A comparison with the specimens before described shows the usually odd isolated cusp to be the homologue of the " paraconid " combined with the anterior odd cuspidule of *Lagopsis*, while the outer cusp of the tripartite anterior lobe is present, also, in p. 2 ; in *P. œningensis* it is generally stouter than the outer cusp (6) of the median lobe, whereas in *P. sardus* the inverse is the rule. In exceptional cases of *P. sardus* I find this outer cusp of the anterior lobe completely isolated, as it is in the deciduous tooth of fig. 5.

A second characteristic feature of the p. 2 of *Prolagus* (figs. 9, 12) is a longitudinal enamel-fold, filled with cement, which, beginning from behind the isolated anterior cusp, proceeds backward to near the hinder margin of the tooth, thus completely dividing the middle lobe into an outer and an inner cusp, and incompletely so the posterior one, on which it also encroaches. The longitudinal arrangement of the elements of this p. 2 of *Prolagus*, in opposition to the transverse arrangement of the posterior teeth, is very striking.

I now proceed to a consideration of the same tooth in the Leporidæ. With reference to $\overline{p}.\ 2$ of *Palæolagus*, Leidy states:—"The anterior four inferior molars [of *Palæolagus*] bear a near resemblance in form and constitution with the corresponding series of *Titanomys cisenociensis*, as represented in pl. 16 of Gervais' Zool. et Pal. Fr." †. Comparing it with the same tooth in *Lepus*, Leidy further says in the original description of *Palæolagus*:—" The first inferior molar is bilobed, and not trilobed as in the latter (*Lepus*)" ‡. In his second memoir the first inferior molar of *Palæolagus* is said to be composed of a double column as in the others, the same tooth in the Hare of a triple column §. Cope

---

* R. Hensel, *l. c.* pl. xvi. fig. 8.

† 'Extinct Mammalian Fauna of Dacota and Nebraska,' p. 333, pl. xxvi. (1869).

‡ Proc. Ac. Philad. p. 89 (1856).

§ Extinct Mamm. Fauna, &c., p. 331.

supplements this description by the following information based upon a great number of remains :—" I am able to show that it is only in the immature state of the first molar that it exhibits a double column, and that in the fully adult animal it consists of a single column with a groove on its external face "*. A more complete description is given on p. 878 :—" There is the merest trace of a posterior lobe "—corresponding to the terminal lobe (t) of *Titanomys*—"at this time, and that speedily disappears. The anterior lobe is subconical, and is entirely surrounded with enamel. By attrition, the two lobes are speedily joined by an isthmus, and for a time the tooth presents an 8-shaped section, which was supposed to be characteristic of the genus. Further protrusion brings to the surface the bottom of the groove of the inner side of the shaft, so that its section remains in adult age something like a **B**." From this description it appears that p. 2 of *Palæolagus Haydeni* is almost exactly like the same tooth in *Titanomys Fontannesi*.

The difference between the p. 2 of *Palæolagus* and *Lepus* is stated by Cope to be as follows:—In the extinct genus the first tooth "consists of one column more or less divided. In *Lepus* this tooth consists of two columns, the anterior of which is grooved again on the external side in the known species." Leidy's description of the Leporine p. 2, as being composed of three lobes or columns, is more accurate. It is quite true that in the adult p. 2 of many Leporidæ appears to be composed of two columns, with an additional antero-external enamel-inflection (see Pl. 37. figs. 13 & 19); but by no means universally so, and, so far as my experience goes, it is never so in the young (Pl. 37. figs. 8, 18, 22, 23).

In the immature p. 2 of *Lepus* s. l. (Pl. 37. figs. 8, 22), as well as in the immature stage of all the other inferior molars of the same, the posterior and the middle-lobe column are completely divided; only in later stages a very narrow isthmus of dentine connects them on the inner side (Pl. 37. figs. 13, 20, 23). The fact of a primary separation into two lobes of the inferior molars of *Lepus* was first announced by Hilgendorf ‡.

The unworn lower p. 2 of the Wild Rabbit (Pl. 37. fig. 8) displays anteriorly the anterior of the three columns completely divided into a smaller outer and a larger inner subconical cusp; this division is brought about by a longitudinal enamel-inflection, which invades part of the middle lobe as well, so that the latter is also divided, though incompletely, into an outer and an inner cusp. (Compare the homologous enamel-inflection of *Prolagus*, fig. 9.)

Passing on to the lower cheek-teeth backward from p. 2, the various stages which I have represented in Pl. 37 show in the lower molars the simple transverse pattern of the two lobes of p. 1 ; m. 2 is a secondary one, as in the upper teeth, though in the inferior molars the original pattern is much more ephemeral, least so in p. 1, which forms a transition between p. 2 and the true molars.

---

* 'The Vertebrata of the Tertiary Formations of the West,' p. 874, pls. 56 & 57 (1883).   † *Op. cit.* p. 870.

‡ " Bestehen die unteren Backzähne anfangs aus zwei getrennten Schmelzlamellen, welche erst später mit einander verwachsen, so dass ein wesentlicher Unterschied zwischen zusammengesetzten und schmelzfaltigen Zähnen der hasenartigen Thiere nicht zu machen ist." Monatsber. d. K. preuss. Akad. d. Wiss. zu Berlin. Sitzg. v. 14. Dec. 1865, p. 673 (1866).

These teeth, as a whole, exhibit in younger stages a greater longitudinal diameter than in the adult; this is notably the case in *Titanomys* (Pl. **37**. figs. 7, 10, 24), and is chiefly due to the greater development and independence of the terminal cusp (*t*).

The youngest mandible of *Titanomys* which I possess is a left ramus of *T. Fontannesi* (Br. Mus. Geol. Dep. M5267 *b*), figured Pl. **37**. fig. 10. P. 2 and m. 3 have dropped out. Flanking the three corners of the alveolus for p. 2 are visible the small alveoli for the roots of deciduous teeth; the anterior and the postero-external seem to belong to d. 2; the postero-internal was presumably occupied by the anterior root of d. 1. P. 1 is still in the socket and had not yet come into use. Both the principal lobes composing this tooth are surrounded by enamel; but the wrinkled central surfaces of the lobes are composed of dentine, with the exception, perhaps, of the summits of some of the wrinkles, which, to judge from their shining appearance, may bear a very slight coating of enamel *. In p. 1 and the true molars of *adult Titanomys Fontannesi*, the enamel bordering appears interrupted in the middle of the anterior margin (Pl. **39**. fig. 6 *a*). Hilgendorf has recorded a similar instance of the absence of the enamel bordering on the inner half of the anterior border in the lower check-teeth, p. 2 excepted, of *Lepus* †. The anterior transverse lobe of p. 1 (fig. 10) still shows traces of having been divided originally into an outer and an inner cusp and of the " paraconid " on its anterior border; vestiges of the latter are visible also on p. 1 of a slightly older individual (fig. 16, of the right side), and on m. 2 of the same right ramus. The terminal cusp *t* ("hypoconulid") is present in both p. 1 and m. 1 of the younger specimen (fig. 10), as well as in p. 1, m. 1, and m. 2 of the second individual (fig. 16), and in p. 1 of a third (fig. 21, right side). In the left ramus, exhibiting the complete series of five check-teeth (fig. 7), *t* is present in all of them. In p. 1 of *T. cisenociensis* (fig. 25) it is remarkably large, although partly fused with the posterior lobe; and it is equally present on the posterior border of m. 1 and m. 2 of the second specimen of *T. cisenociensis* (fig. 24); so that, contrary to what has been stated by former writers, the cuspidule in question may be present in all the four anterior check-teeth of this species.

Passing on to the recent representatives of the family, it may be seen from fig. 22 (Pl. **37**.), of an immature *Caprolagus hispidus*, that p. 1 nearly approaches p. 2 in its anterior complication. The two principal lobes are not yet connected on the inner side by a dentinal isthmus, but are merely held together by cement; the anterior lobe is distinctly composed of an outer and an inner cusp, the latter being more pointed and slightly higher than the former. The anterior border of the tooth presents two minor cusps, an outer and an inner, the median odd cusp of p. 2 being absent. Both the lobes show a very marked wrinkling of their surface. As in p. 2, *t* is apparent on the posterior margin of the second lobe.

Two very distinct minor cusps are likewise visible on the anterior border of p. 1 of the

---

* We have here an instance similar to that recorded by Hensel in *Mus decumanus, rattus, musculus, sylvaticus, agrarius,* and *minutus,* where in perfectly unworn molars " überzieht der Schmelz die Hocker der Zahnkrone niemals vollständig, sondern lässt an den Spitzen das Zahnbein frei hervortreten." Zeitschr. d. deutsch. geol. Ges. viii. pp. 283, 284, pl. xiii. figs. 2, 3 (1856).)

† Sitzungsber. Ges. naturf. Freunde zu Berlin, 14 Jan. 1884, p. 23.

Rabbit (fig. 8); the first lobe of the same is mainly composed of an outer and an inner cusp, separated by a median hollow ; the second lobe is wrinkled as in p. 2 of the same species. The minor cusps, though less distinct, are visible also in m. 1 and m. 2 of the Rabbit, in m. 2 almost vanishing. I have noted their presence in the true molars of young specimens of other species as well (*Lepus europæus*, *Lepus* sp. from China, *Sylvilagus brasiliensis*): *l* is generally present in unworn deciduous teeth, in premolars, and in molars of several Leporidæ.

To sum up the above as regards the lower check-teeth, p. 2—m. 2. An original arrangement into outer and inner cusps, separated by a median longitudinal valley, is traceable in the lower molars of Lagomorpha generally. It is more distinct in the anterior check-teeth, and persists throughout life in p. 2 of most genera in both families ; it is less distinct, though perfectly perceptible, in true molars, in which it very soon disappears by wear, being replaced by the transverse arrangement. In p. 2 we have to distinguish between an older complication and secondary additions ; the increase in the plication alone is present in the posterior check-teeth, the anterior cusp not. On comparing adult stages of p. 2 of *Titanomys* with the corresponding tooth of all other Lagomorpha which, on the whole, are more recent forms, the latter appear to be more complicated ; but in young stages p. 2 of *Titanomys Fontannesi* presents also a complicated appearance. This cannot be an incipient complication, for that part of the shaft of the tooth which is situated on the opposite end of the pulp-cavity is, as a matter of course, always the oldest. Hilgendorf has found the interruption of the enamel border on the inner side also of lower molars of *Lepus* [*], a fact which points towards a degeneration of this part of the tooth, and would seem to call for a compensatory increase on its outer side. However, I am not aware of a perceptible additional increase on the outer side of lower molars of more recent forms, as compared with older ones [†].

Upper molars are more progressive than lower as concerning occasional additions. An ingenious explanation of this general occurrence is given by Winge in the following remark:—" The explanation of the maxillary teeth making a larger increase than those of the lower jaw is in all likelihood the following : they are placed in an unmovable bone, where the conditions for nourishment are more favourable than in the comparatively slender and movable mandible " [‡]. In our special case an increase of the lower molars in the transverse direction can be the more dispensed with, since in the Leporidæ the movement of the jaws is chiefly lateral. This will not be denied by any one who has ever examined the shape of their glenoid cavity or watched a Rabbit or Hare chewing. Moreover, the dentine of both upper and lower check-teeth shows unmistakable signs of this movement, in the presence of transverse striæ, due to the action of the transverse enamel crest of the opposite tooth.

It remains to discuss in some detail the last molar, m. 3, about which very divergent views have been put forward.

---

[*] *Op. cit.* p. 23.

[†] Neither am I aware of lacunæ on the *internal* enamel bordering of any Lagomyidæ ; but I must add that no sections were made.

[‡] Vidensk. Meddelelser naturhist. Forening i Kjøbenhavn f. Aar. 1882, p. 17 (1883).

Fig. 7, Pl. 37, shows this tooth in place in a left mandibular ramus of *Titanomys Fontannesi*. It is not a simple cylinder, as in *Lagopsis* and *Lagomys*, but is composed of two lobes, a larger anterior one and a small posterior, attached to the former in the same manner as in the anterior molars the terminal cusp (*t*) is attached to the lobe preceding it, viz. separated from it by cement, only in the upper part. For this reason, and because the anterior lobe of m. 3 shows traces of greater complication, I homologize the posterior lobe of this tooth with *t* of the anterior molars; the anterior lobe of m. 3 would then represent *both* the principal lobes of the anterior molars.

When discussing the tooth-formula of *Titanomys*, allusion was made to Filhol's suggestion that the terminal cusp of m. 2 of *T. visenoviensis* might be the representative of m. 3 of the recent *Lagomys*, in the specimens of the former where this is missing. "Si cette opinion est juste, on pourrait en tirer comme conclusion qu'à un certain moment, sur les animaux voisins des *Lagomys*, il y a une tendance à la simplification du système dentaire, d'abord par la fusion de la dernière dent avec l'avant-dernière, et ensuite par la tendance à la disparition de cet élément soudé " [*]. Filhol here ignores the circumstance that all the anterior teeth have this "troisième lobe" as well, while in their case we have not at our disposal an occasional small isolated tooth to suggest a fusion theory. Besides, as was said before, this theory may be at once disposed of by a glance at our fig. 7, showing m. 2 with a well-developed terminal cusp (*t*), m. 3, the supposed homologue of this latter, being likewise present. Other figures also (figs. 10, 16) show m. 2 with the terminal cusp, together with the alveolus of m. 3.

As will be seen further on, Schlosser seems to incline to the opinion that the presence of a terminal cusp in m. 2 of *T. visenoviensis* is an indication of m. 3 having become fused to m. 2; for he says that m. 3 of *Lagopsis cerus* may be the analogue of the terminal cusp (*t*) in m. 2 of *Titanomys* [†]. It is, however, difficult to make out what meaning he wishes to attach to this vague term "Analogon".

Lagopsis.—The type-specimen. Hensel's *Lagomys cerus* [‡], has five lower cheek-teeth, the last being a small cylindriform tooth, precisely as in the recent *Lagomys*, to which *Lagopsis* is closely related. The tooth in question was not complete in Hensel's specimen, but a fragment seems to have remained inside the alveolus; else he would have presumably used the term "ausgefallen," whereas he says, speaking of the condition of this tooth, that it is broken away (" weggebrochen ").

Three more or less complete mandibular rami, from Deggenhausen, Elgg, and Hohenhöven respectively, are mentioned by H. v. Meyer, and drawings of their teeth, found among H. v. Meyer's MSS. have been reproduced by Schlosser §. They show an agreement in their p. 2 with Hensel's *Lagomys cerus*, and Schlosser therefore concludes ||, rightly, I think, that they are of the same species. He further deems it not improbable ¶ that *Lagomys oeningensis*, H. v. Mey., from Œningen may be identical with *Lagomys*

---

* Ann. Sc. Géol. x. p. 28 (1879).

† ' Nager des europ. Tertiars.' p. 32 (1884).

‡ Zeitschr. d. deutsch. geol. Ges. 1856, p. 688, pl. xvi. figs. 12, 13.

§ *Op. cit.* p. 31, pl. viii. figs. 40, 46, 49.

| *Op. cit.* pp. 31, 32.

¶ *Op. cit.* p. 32.

verus, Hens." That this is true with regard to the Œningen specimen in the British Museum has been shown on p. 462. I can affirm the same for the Seyfried specimen * at present in the Constance Gymnasium, where I examined it and found it to have the characteristic $\overline{p.}2$ of *Lagopsis verus*. With regard to the Carlsruhe specimen †, since the shape of its $\overline{p.}2$ cannot be clearly made out from H. v. Meyer's figures and description, the true position of this "*L. œningensis*, H. v. Mey.," cannot be satisfactorily determined for the present. It might quite as well be a *Titanomys Fontannesi*. In the former, as well as in the specimens from Deggenhausen, Elgg, and Hohenhöven, no last molar ($\overline{m.}3$) could be seen ; as, however, this tooth is very caducous, its absence in the fossils is not in the least conclusive ; it may have dropped out and the alveolus been filled with matrix. Nor does Schlosser attach any great weight to the absence of this small tooth in the three specimens drawn in H. v. Meyer's MSS. ; this, however, for reasons with which I completely disagree. "Auf das Fehlen des letzten einfachen Backzahnes bei den drei von H. v. Meyer gezeichneten Exemplaren darf wohl nicht allzuviel Gewicht gelegt werden. Es ist nicht unmöglich, dass auch hier, wie bei *Titanomys visenoviensis*, im normalen Kiefer nur 3 zweilobige Molaren vorhanden sind, und dass daher der stiftförmige m. 4 " (meaning m. 3) " des Hensel'schen Originales als Analogon des bei *T. visenoviensis* abnorm vorkommenden Lobus des m. 3 " (meaning m. 2) "betrachet werden muss." ‡

This whole statement is somewhat vague : the author seems to assume (1) that in *T. visenoriensis* both the m. 3 and the third lobe (*l* in my figures) of m. 2 occur only abnormally ; (2) that in "*Lagomys verus*" the presence of m. 3 is equally an abnormal occurrence. From these two assumptions the inference is drawn that m. 3 in the type of *Lagomys verus* is the analogue of the equally abnormal third lobe in m. 2 of *T. visenoviensis*. Schlosser concludes by saying that he is almost inclined to consider the presence of $\overline{m.}3$ as a juvenile character, and that this tooth is caducous (hinfällig). This is very probably true with regard to *T. visenoriensis*, and I have myself suggested it in the preceding pages. But it is decidedly erroneous with regard to $\overline{m.}3$ of *Lagopsis verus*, as are all the other suggestions tentatively put forward in the passage quoted. With regard to *T. visenoriensis*, the matter has been fully discussed above. As to the m. 3 of *Lagopsis verus*, in all my specimens from La Grive-Saint-Alban, either the tooth itself or its very distinct alveolus is present (Pl. 37. figs. 14, 26). Depéret, too, has before figured a mandibular ramus of *Lagopsis verus* from the same locality, showing the m. 3 § ; and Biedermann has described this same tooth in specimens from Elgg.

Prolagus.—There is no third inferior true molar, $\overline{m.}3$, in this genus ; m. 2 is composed of three lobes, the posterior connected with the middle one by cement, in the same way as the latter is with the anterior one. From this circumstance Pomel concluded— ust as Filhol has in the case of *Titanomys* —that in *Prolagus* m. 3 had become fused with m. 2. Of the *Prolagus oeningensis* of Sansan, he says:—"Ceux de Sansan diffèrent

* H. v. Meyer, "Fossile Säugethiere, etc., von Œningen." Fauna d. Vorwelt, p. 6. pl. iii. fig. 1 (1845).
† *Ib.* pl. ii. fig. 1.
‡ *Op. cit.* p. 32.
§ Arch. Mus. Lyon, iv. p. 164, pl. xiii. figs. 16, 16 *a* (1887).

encore, comme sous-genre, par la dernière molaire inférieure, qui a trois prismes par réunion de la cinquième molaire à la quatrième "[*]. Fraas holds the same opinion [†].

This theory would at first sight seem to be supported by what Depéret has found in the *Prolagus* of Roussillon. He figures two mandibular rami [‡], in one [§] of which he records five check-teeth, in the other [||] only four; and he goes on to say :— "Cette différence est moins importante qu'elle ne peut sembler au premier abord ; elle tient simplement à ce que le dernier prisme d'émail de la série dentaire est soudé au prisme précédent de la quatrième molaire dans l'une de ces mandibules, tandis que ce même prisme libre constitue une cinquième molaire dans la fig. 29. Cette soudure, qui se fait d'ailleurs uniquement par l'intermédiaire d'une certaine quantité de cément, ne me paraît pas avoir l'importance qu'on lui a attribuée pour la distinction des deux genres *Lagomys* et *Prolagus*, puisqu'elle est variable suivant les sujets dans le petit Léporidé de Roussillon "[•].

I agree with Prof. Depéret that this difference has no great importance in the Roussillon jaws, though not for the reasons adduced, for I apprehend he is mistaken when he institutes comparisons with *Lagomys*, and considers that the isolated prism of his fig. 29 "constitue une cinquième molaire." H. v. Meyer met with similar occurrences among twenty mandibular rami of *Prolagus œningensis* (Kön.) from Steinheim, and refers to them in the following words :—" In some instances one might be induced to believe that the posterior of the three prisms constituting the last molar is separated, so that the creature would have the character of *Lagomys* "; but he judiciously adds :—" On closer examination, however, it can be seen that the posterior prism is included in the alveolus of the rest of the tooth, so that it evidently is part of the latter " (" dass das hinterste Prisma nicht durch die Alveole von dem übrigen Zahn abgeschlossen ist, zu dem es daher offenbar noch gehört) "[**]. Numerous mandibular rami of the *Prolagus œningensis* from La Grive have passed through my hands, as well as from 600 to 700 of *P. sardus* from the Corsican and Sardinian ossiferous breccias and caves. Not unfrequently I found the third prism of m. 2 separated from the rest of the tooth; but by the criterion established by H. v. Meyer there could never be a doubt as to the interpretation, which invariably was that, either by fracture or by the weathered condition of the cement, the last prism had been separated from $\overline{m. 2}$; as are likewise, though more rarely, separated from each other the two prisms of the anterior teeth. I do not doubt for a moment that the same explanation will hold good in the case of the Roussillon specimens. In *Prolagus* each of the prisms has its alveolar niche formed by two partial septa starting from the outer and inner alveolar border ; but these must not be confused with the complete septum separating one alveolus from the other.

I consider the third prism of m. 2 of *Prolagus* to be the homologue of *t* of the

* Cat. méth. et descr. Vert. foss. du Bassin de la Loire et de l'Allier, p. 43 (1853).
† Wurttemb. naturw. Jahresh. xxvi. p. 170 (1870).
‡ " Anim. plioc. du Roussillon," Mém. Soc. Géol. France, i. p. 57, pl. iv. (1890).
§ *Op. cit.* pl. iv. figs. 29, 29 a.
|| *Op. cit.* pl. iv. figs. 28, 28 a.　　• *Op. cit.* p. 57.　　** Neues Jahrb. 1865. p. 813.

Titanomys-teeth; and that m. 3 having been lost in the former genus by some means or other, the terminal cusp of m. 2 has become enlarged in compensation. We have numerous analogies for similar occurrences, but we have none for the ever-recurring theories of fusion between tooth and tooth, which on closer examination always break down. This notwithstanding, we shall still hear of them, since they yield the explanation which lies nearest at hand.

Again, although *Prolagus* presents in its molars, at least in the upper ones, more primitive characters than *Lagopsis* and *Lagomys*, it cannot be considered to be the direct ancestor of these; for it cannot be surmised that a tooth—m. 3—after having been lost, reappears in a later genus. Hilgendorf regards m. 3 of *Lepus* as a recent acquisition, for he terms it "phylogenetisch der jüngste (Zahn)" *; presumably for the same reason for which he considers the maximum of enamel-plication observed by him in upper incisors (of "*Lepus mexicanus*") to be "phylogenetisch ein Extrem" †, because there is no trace of it "bei den fossilen Leporiden-Gattungen (*Myolagus*)." There is no good reason for considering the Miocene *Prolagus* (*Myolagus*) in the ancestral line of *Lepus*, simply because no true Leporidæ have been found in the European Miocene; nor in inferring from the various primitive characters of *Prolagus* that the absence of m̅. 3 is a primitive character as well. Besides, Hilgendorf does not take into consideration the fact that *Lagopsis* and *Titanomys*, both of which are contemporaneous with and even partly (*T. visenoviensis*) older than *Prolagus*, possess a m̅. 3. I presume that, for similar reasons, Hilgendorf would consider the m. 3 of *Lepus* a recent acquisition also; and here we must remember that the Oligocene *Palæolagus* has both m. 3 and m̅. 3.

Noack describes the last lower molar of young *Lepus saxatilis* as composed of two antero-posteriorly placed cusps, which seem ("scheinbar") to be separate, but at any rate ("jedenfalls") are only loosely connected, which makes it doubtful whether they ever coalesce to form a compact tooth. This conformation of m. 3 is in the author's opinion a sufficient justification for the following generalization: "Jedenfalls ist im Unterkiefer von *L. saxatilis* noch die Tendenz zu 6 Backenzähnen vorhanden." ‡ Why not, while we are at it, towards eight?—since it is stated immediately afterwards that the same partitioning of the two lobes is also visible in two of the anterior molars. The

---

* Sitzungsber. d. Ges. naturf. Freunde Berlin, Sitzung v. 15. Januar 1884, p. 23.

† *Op. cit.* p. 20.

‡ Th. Noack, "Neue Beiträge zur Kenntniss d. Säugethier-Fauna von Ostafrika," Zool. Jahrb. Abth. f. Syst. etc. vii. p. 545 (1893). The writer of this pamphlet has examined numerous dentitions of fœtal and young Rabbits, and "*L. vulgaris*" (meaning *L. europæus*), and finds among other things in their cheek-teeth cusps which are absent in the adult. So far, good. Apart from this, his descriptions and generalizations show on almost every line that he has approached this difficult subject without sufficient scientific training. Hilgendorf's short sentence of 1865: "Die oberen Backzähne junger Hasen sind mit einer halbmondförmigen Schmelzröhre versehen, wodurch ein Übergang zu dem fossilen *Myolagus* gebildet wird,"—is of infinitely higher scientific value than the pages filled with laborious descriptions in the paper quoted. If the author had taken Hilgendorf's words as a starting-point and a guide in the investigation of upper leporine cheek-teeth, he might have been able to do some useful work. He knows about tritubercular teeth; he also seems to be aware that on one occasion the molars of lagomorphous Rodents have been compared with those of diprotodont Marsupials, and that

numerous juvenile dentitions which were at the author's disposal might have shown him that the separation of the two lobes is characteristic of young stages in the inferior cheek-teeth of *Lepus* generally.

THE BONY PALATE IN THE LAGOMORPHINE SKULL.

The greatly reduced bony palate is considered to be one of the characteristic features in the skull of Lagomorpha. At first sight the only difference in this respect between Leporidæ and Lagomyidæ appears to be that in the latter family the palatal bridge is shorter than in Leporidæ. On investigating the matter more closely, however, it may be seen that in Leporidæ the bony palate is shortest in the genus *Lepus* s. str., viz. in those forms which are most specialized for running and leaping; and that the shortness is principally due to a reduction in length of the os palatinum. In Lagomyidæ, on the contrary (Pl. **39.** figs. 34, 36, " *p* "), the latter bone is comparatively elongate, while the part of the bony palate formed by the maxillaries (*m*) is greatly reduced, so that in some cases the latter do not even join in the middle line anteriorly, the middle of the anterior margin of the palatal bridge being formed by the palatine bones. As seen from the figures, *Prolagus* (fig. 36) is in this respect scarcely different from *Lagomys* (fig. 34).

It might, *a priori*, be expected that this specialization of the Lagomorpha will be reduced to a minimum, in other words that the bony palate will be longest, in the oldest members of the group, and this is in fact so. Cope describes this part of the skull of *Palæolagus* as follows :—" The palatine bones are flat and occupy more than half the palate between the molars. Their common suture is at least as long as that of the maxillaries, and extends as far forward as the posterior border of the second molar. From this point the anterior suture extends to the posterior border of the third molar. The palatal notch is rectangular, and is not wider than the palatine bone on each side of it." *

some phylogenetic speculation has been based thereon. The author avails himself of these two types, the tritubercular and the diprotodont, in tracing two primitive types in the teeth of one species, *Lepus saxatilis* ; the anterior upper cheek-tooth is referred to the tritubercular type ; the conformation of the two anterior lower teeth, on the other hand, " decidedly suggests the molars of Kangaroos and Wombats, and makes it probable that the ancestors of the Lagomorpha were Marsupials, holding about the middle between *Phascolomys* and *Lagorchestes*" (p. 545). By the cheek-teeth of its ripe embryo, the Wild Rabbit is far removed from *Lepus europæus* (p. 553); and the cheek-teeth of the latter were evolved from the tritubercular type (p. 551). The rabbit's skull approaches the Marsupial type (p. 551). The author seems to be unaware of the existence of deciduous cheek-teeth in the Leporidæ. On p. 549, the anterior of the upper cheek-teeth is twice termed p. 1. Supposing that we have really to do with a premolar, the anterior premolar in the upper series would be p. 3, according to Hensel's mode of writing, adopted by the present writer, or p. 2, according to the usual custom, but under no circumstances p. 1. Considering, however, that the two teeth referred to by Prof. Noack belong, the one to a mature, the other to an unripe embryo of *L. europæus*, in which species the tooth-change takes place only some time after birth, the alleged p. 1 is in reality a d. 3 (d. 2 of authors). On pp. 544 and 545 the remarkable circumstance is noted that in the half-grown *L. saxatilis* the second and third anterior upper cheek-teeth are more retarded in their development than the same teeth in embryos of *L. europæus*. The very obvious explanation is that those of the former species are premolars, those of the latter deciduous teeth.

* E. D. Cope, 'The Vertebrata of the Tertiary Formations of the West,' i. p. 875 (1876) pl. lxvi. figs. 1, 4 (1883).

The only known palate of *Titanomys* is that figured by Filhol *, which too is elongate. According to him †, the length of the palatal bridge in *Lagomys* and *Titanomys* respectively is as follows :—

	millim.
Lagomys tibetanus	0·002.
Lagomys ogotona	0·0015.
Titanomys risenoviensis	0·0015.

The suture between the palatines and maxillaries is not shown in the figure of *Titanomys*. Thanks to the kindness of Mons. M. Boule, I have been able to examine the original in the Paris Museum, and can state that in this oldest member of the Lagomyidæ the family character is already very evident in the reduction of the maxillaries, inasmuch as the palatines occupy the anterior margin of the bridge in the middle line, the two maxillaries not joining each other. The difference in the length of the palatal bridge between *Titanomys* on the one side, and *Lagomys* (with *Prolagus*) on the other, is therefore wholly due to the greater elongation of the former's palatine. In *Palæolagus* ‡ both bones are lengthened, as compared with other Leporidæ, and especially with the most modernized species of the family. The anterior palatal notch formed by the maxillaries extends forward slightly beyond the anterior margin of p. 3, as it does in *Nesolagus Netscheri* (Pl. 39. fig. 38), which is one of the most primitive of recent Leporidæ. The posterior palatal notch of *Palæolagus* reaches as far backward as a line uniting the middle of the alveoli of m. 1. Besides, the horizontal portion of the ossa palatina is also transversally much less reduced than in most of the recent Leporidæ, the breadth of the posterior palatal notch being approximately equal to half the breadth of the space between it and the alveoli. While in this latter character *Palæolagus* converges towards the Lagomyidæ, or rather goes beyond them—for, to judge from the figures, the palatal notch of *Palæolagus* is considerably narrower than even in *Titanomys*—it is thoroughly leporine with regard to the part which the maxillaries take in the formation of the bony palate.

Those among recent Leporidæ which, on account of their several primitive characters, may be placed in a separate section (Caprolagus-group), as opposed to *Lepus* s. str., are more primitive also in the character of the greater antero-posterior length of the palatal plates of the palatine and maxillary bones, as may be judged from various instances figured in Pl. 39. Fig. 32 represents the palate of *Caprolagus hispidus* (Pears.) ; fig. 33, of *Sylvilagus* (*Romerolagus*) *Nelsoni*; fig. 37, the same part of *Oryctolagus crassicaudatus* (Geoffr.) ; fig. 38 that, already mentioned, of *Nesolagus Netscheri* of Sumatra. It is well known that the bony palate of the Rabbit, of which a figure is not given here §, has a greater longitudinal extension than in the Common Hare and that its palatal notch is narrower; both these characters are much more pronounced in the young. Fig. 35

---

* H. Filhol, " Étude des Mammifères fossiles de Saint-Gérand-le-Puy (Allier)," Ann. Sc. Géol. x. pl. 3, fig. 16 (1879).   † *Op. cit.* p. 31.

‡ Cope, *op. cit.* pl. lxvi. figs. 1, 4.

§ Excellent lower views of skulls of the Rabbit, side by side with those of *Lepus europæus*, have been figured by H. v. Nathusius (' Über die sogenannten Leporiden,' pl. ii. 1876).

(Pl. **39**) represents these parts of a young *Sylvilagus brasiliensis* (Linn.), which closely resembles *Palæolagus* in the great antero-posterior extension of both the palatine and the maxillary bones and in the very narrow palatal notch, both coming near to the normal condition of Mammals.

As might have been expected, the Pliocene *Lepus valdarnensis*, Weith., also presents a more normal palatal region than the various specialized species of *Lepus*, and may for this reason alone be assigned to the *Caprolagus* section. The anterior and posterior palatal notches are much narrower than in *L. europæus*, and the whole of the bony palatal bridge is considerably longer ; this being especially due to the elongation of the maxillaries *.

The greater reduction of the palatal plate of the maxillary bone in Lagomyidæ, as compared with Leporidæ, might seem to be due to the greater backward prolongation of the foramina incisiva in the first-named family. On closer examination, however, it becomes evident that in reality we have to do with a fusion of two originally separated vacuities, viz. the true foramina incisiva, and a sort of palatal fontanelle behind them. In *Lagomys*, the premaxillæ generally, though not in all the species, join in the middle line between the foramina incisiva and the fontanelle behind them; in Leporidæ, the confluence of the two fissures has generally, but not always, become complete. An approach to Lagomyidæ (fig. 36) is given by the bottle-shaped appearance of the "foramina incisiva" which Bangs considers to be characteristic of "*Lepus sylvaticus transitionalis*"† —the same occurs also in other American Leporidæ—and which is but the remnant of the original separation of the true foramina incisiva from the palatal fontanelle. I therefore do not think that Winge is right, when he assumes that the separation of the two openings is a secondary character in *Lagomys*, brought about by the new formation of a bony plate ‡. Judging from Cope's figure §, the fusion of both openings seems to have already taken place in *Palæolagus*. But if we judge from recent forms, in which the premaxillæ are very thin in this region, it appears probable that the apparent fusion in the figured palate of *Palæolagus* is due to the defective preservation of the premaxillæ in the figured specimen.

On the Limb-Skeleton of Lagomorpha.

There is a great difference between the Lagomyidæ and Leporidæ, and between the various members of the latter, in the absolute length of the fore and hind limbs, and in their relative length, compared with each other. The differences, moreover, are not only in size; and it is the antebrachium which in the first place presents notable divergences in the different groups. Even for systematic purposes it will be necessary henceforth to take into consideration these, as well as other, parts of the skeleton; and we cannot content ourselves with such general statements as " hind limbs longer than the fore limbs," and " hind limbs and fore limbs subequal."

---

* A. Weithofer, in Jahrb. k.-k. geol. Reichsanstalt, Bd. xxxix. p. 80 (1889).

† Proc. Bost. Soc. Nat. Hist. xxvi. p. 407 (1895).

‡ H. Winge, 'Jordfundne og nulevende Gnavere,' &c., *l. c.* p. 113 : " Forskjellen fra Haren er kun, at det egenlige *F. incisivum* er afskilt ved en nyopstaaet, ikke altid fuldstændig Benbro."   § *Op. cit.* pl. lxvi. fig. 1.

In comparing the characters of the common Hare (*L. europæus*) with those of the domesticated Rabbit, Nathusius enters into full particulars of the differences presented by the antebrachium, summing them up in the following statements:—

Hare.	Rabbit.
Ulna weaker than the radius, situated behind the latter.	Ulna stronger than the radius, situated laterally.

In relation to the basilar length of the skull and the length of the vertebral column, the anterior and posterior limbs are in their totality, as well as in their different parts, *longer* in the Hare, *shorter* in the Rabbit.

Hare.	Rabbit.
Humerus longer than antebrachium.	Humerus and antebrachium subequal in length.

Length of the antebrachium as compared with the tibia :—

Hare.	Rabbit.
Antebrachium shorter than the tibia by about one-fourth its length.	Antebrachium shorter than the tibia by one-half its length [*].

With regard to the remarkable differences in the antebrachium of the two animals, the writer concludes that they are doubtless associated with their different habits, the Rabbit burrowing and the Hare living above-ground [†]. Put in this general way, the conclusion is undoubtedly true. Nathusius, however, does not seem to have been aware that the difference is chiefly due to the specialization of the *Hare's* fore-leg, which specialization is nothing else than the beginning of the process carried much further in the modern swift-footed Ungulates. It therefore remains to be seen how far, if at all, the structure of the Rabbit's antebrachium is a consequence of its burrowing propensities,—an adaptation to them. For neither from what we know of its habits, nor from the structure of its fore-limb, can the Rabbit be considered to be a truly fossorial Mammal, as is, *e. g.*, the Mole, or, among Rodentia, the genera *Geomys*, *Spalax*, and *Siphneus*.

In districts where the Rabbit finds burrowing in the ground too hard a task, it manages to do without it [‡]; as it sometimes does, perhaps, for other unknown reasons.

---

[*] H. v. Nathusius, ' Über die sogenannten Leporiden.' pp. 17, 31-33, 67, figs. 2-5 (p. 32) 1876.

[†] *Op. cit.* p. 33.

[‡] W. Thompson states (Proc. Zool. Soc. London, part v. p. 52, 1837) that in the North of Ireland persons who take Rabbits make a distinction between the *Burrow-Rabbit* and the *Bush-Rabbit*, and that the latter is so designated in consequence of having a "form like the Hare, and which is generally placed in bushes or underwood." The Rev. G. T. Dawson, speaking of the Wild Rabbit, says :—" There is a variety . . . which never burrows in the ground, but lies beneath bushes, or among the herbage of hedges or woods, and is called by the common people of that part of Hertfordshire which borders upon Bedfordshire the *Bush-Rabbit*, and in the northern parts of the same county the *Stub-Rabbit* . . . . A non-burrowing Rabbit may, in its distress, scramble into a hole, or burrow, if there happens to be one in its way, in which to die in secrecy ; but, as far as my own observation extends, I never remember one

One of my principal reasons for separating a certain number of Leporidæ, under the designation of *Caprolagus*, from the swift-footed *Lepus* (figs. XXV–XXVIII), is the

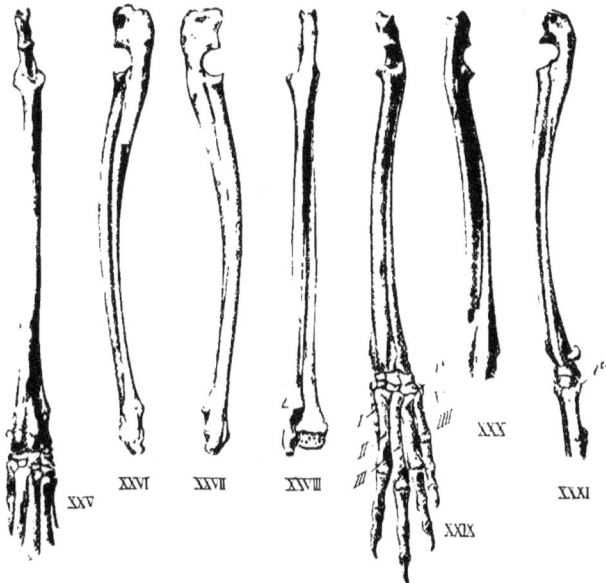

Figs. XXV–XXVIII. Left antebrachium of *Lepus timidus*, Linn. (*L. variabilis*, Pall.). ½ reduced. XXV, front view XXVI, ulnar (external) view ; XXVII, radial view ; XXVIII, posterior view.

Figs. XXIX–XXXI. Left fore-limb of *Sylvilagus brasiliensis* (Linn.), nat. size. XXIX, front view. I–V = first to fifth metacarpals. *c* = carpale 5 (vesalianum) ; XXX, radial view ; XXXI, ulnar view.

structure of the antebrachium ; but of several of the former it is expressly stated that they do not burrow at all, or at least that they are not habitual burrowers. I have thought it would be instructive for my present purpose to record the observed facts of the physiology of the organs of locomotion of various Lagomorpha, by collecting as much information as is available to me.

of the *bush-rabbits* running to ground, even when wounded, and certainly it is contrary to its habits to do so under different circumstances " (' Zoologist,' iii. p. 903, 1845). In W. Thompson's ' Natural History of Ireland ' (vol. iv. p. 30, 1856), his former statement is repeated, and strengthened on the authority of Dr. R. Ball, " who has long been aware of the difference of habit and appearance between burrow- and bush-rabbits in the County of Cork." In Bell's ' History of British Quadrupeds ' (2nd ed. pp. 344, 345, 1874) it is reported that " on moors, where the soil is wet, Rabbits often refrain from burrowing, and content themselves with runs and galleries formed in the long and matted heather and herbage. In more than one instance we have known a family to take possession of a hollow tree and ascend its inclined and decayed trunk for some distance." In comment on this, Prof. Howes has drawn my attention to the fact that the Oriental Black-necked Hare (*L. nigricollis*) habitually resorts to the hollows in trees when pursued, and that while the European Rabbit may bring forth its young above-ground (' Zoologist,' ser. 3, vol. i. p. 18) the Hare may do so in a burrow.

Of *Oryctolagus crassicaudatus*, which, in the conformation of its antebrachium (text-figs. XXXII–XXXV), is almost identical with *O. cuniculus* (Pl. **38**. fig. 30), Smith says in a general way that it inhabits "rocky situations" in South Africa, and that "its manners connect it closely with the Rabbit." * Alexander Whyte describes the same species in his journey through the high-lying country in the North Nyasa district, and he

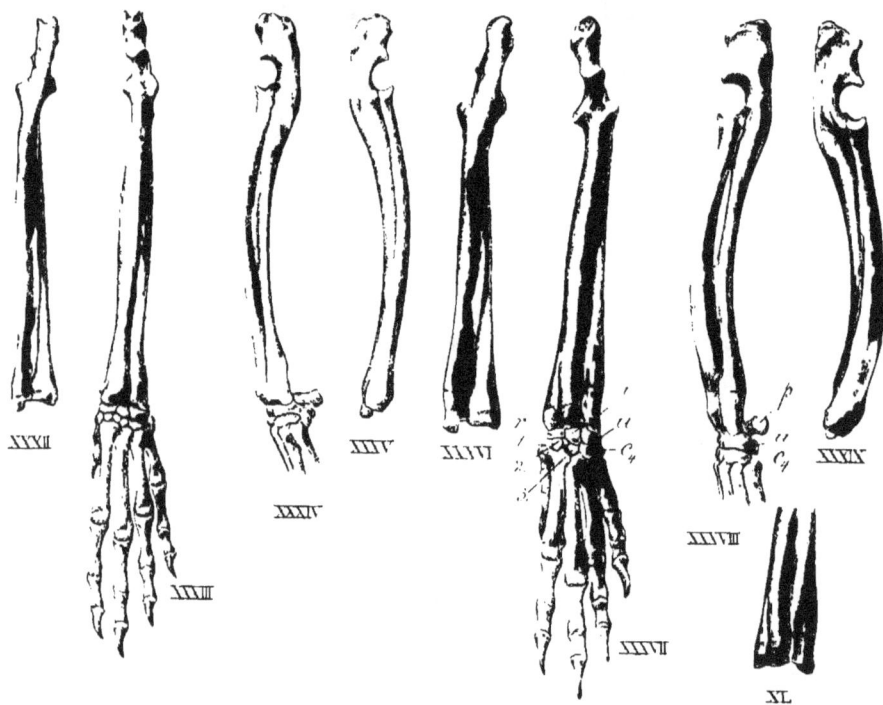

Figs. XXXII–XXXV. *Oryctolagus crassicaudatus* (Geoffr.).—Left fore-limb, nat. size. XXXII, posterior view ; XXXIII, front view ; XXXIV, ulnar view ; XXXV, radial view.

Figs. XXXVI–XL. *Caprolagus hispidus* (Pears.).—Left fore-limb, nat. size. XXXVI, posterior view ; XXXVII, front view: r=radiale, i=intermedium (lunar), u=ulnare. 1–3=carpalia 1–3. C_4=carpale 4 (hamatum). XXXVIII, ulnar view ; p=pisiform ; XXXIX, radial view : XL, front view of antebrachium, proximal end.

too compares it with the Rabbit †. But nowhere have I found it expressly stated that this species is burrowing; the rocky "situations" and "places" to which, according to both observers, it is confined, certainly would not favour burrowing propensities.

* A. Smith, in S. Afr. Quart. Journ. vol. ii. p. 87 (1833) (sub "*Lepus rupestris*").

† "Perhaps the most interesting mammal we secured was the hare of the plateau, and which might well be termed a 'rock-rabbit.' . . . It is very local and peculiar in its habits, confining itself to the highest and most rocky places on the plateau. On this account we found it most difficult to procure good specimens. It kept dodging about the granite boulders, and we seldom got a shot until it was quite close on to us. . . . It was never found out in the open . . . . ." (British Central Africa Gazette, 15th Oct. 1895 to 1st Feb. 1896, p. 22.)

Of the "*Lepus brasiliensis*" of Paraguay, whose fore-limbs (text-figs. XXIX–XXXI) much resemble those of the Rabbit, D'Azara states expressly that it is not a burrowing animal *, and the same is confirmed by Rengger †.

About the habits of *Sylvilagus sylvaticus*, the "Grey Rabbit" of the United States, we know from Bachman that "though it digs no burrows in a state of nature, yet when confined it is capable of digging to the depth of a foot or more under a wall in order to effect its escape" ‡. *S. artemisiae*, closely related to *S. sylvaticus*, is described by Clark as burrowing §.

Special recognition is due to the following graphic description by Coues of the locomotion of three different groups of Hares, viz. the Marsh-Hare (*S. palustris*), the "Wood-Rabbit" (*S. sylvaticus*), and the "Jackass Hares" (*L. callotis*). Comparing in the first place the two former, he says:—"The Marsh-Rabbit ... looks smaller, although actual measurement does not show any very decided difference in size. This deceptive appearance is owing to the different gait ... The animal's gait ... is a direct conse-quence of the comparative shortness of its legs—of the hinder ones particularly ... The animal's general configuration is more squat and bunchy; it seems to run with its body nearer the ground ‖, scuttles along with shorter, quicker steps, more constrained and spasmodic, moving by jerks, as it were; and has little or nothing of the free bouncing movements that mark the progress of the Wood-Rabbit. In these respects the last-named species is exactly intermediate between the Marsh-Rabbit and the large "Jackass" Hares (*Lepus callotis*) of the West, in which length of stride, height of bound, and general freedom of swinging gait reach an extreme. These Western Hares are the swiftest of their tribe in this country, and the Marsh-Rabbit is just the opposite As attested by all observers, the speed of the latter is appreciably less than that of even

---

* "Il ne fouille point de terriers, quoiqu'on dise, qu'étant poursuivi, il se cache sous des troncs pourris et entre les débris des végétaux." ('Essais sur l'Hist. nat. des Quadrupèdes de la Province du Paraguay,' ii. p. 58, 1801).

† J. R. Rengger, 'Naturgeschichte der Säugethiere von Paraguay,' 1830.—"Höhlen oder unterirdische Gänge gräbt es keine" (p. 248). "Sein erster Lauf ist schnell; er hält aber nicht lange aus und wird bald von den Hunden eingeholt" (p. 250).

‡ Journ. Acad. Nat. Sci. Philad. vol. vii. p. 335 (1837). The following statement as to the feeble endurance in running of *S. sylvaticus* is almost identical with what Rengger says of *S. brasiliensis*:—"Although it runs with considerable swiftness for a short distance, yet it soon becomes wearied, and an active dog would overtake it, did it not retreat into some hole of the earth, into heaps of logs or stones, or into a tree with a hole near its roots. ... In the Northern States, where the burrows of the Maryland marmot and skunk are numerous, this hare retreats to their holes" (*op. cit.* p. 328).

§ "Wherever the thorny clumps of chapparal and the loose sandy soil afford protection to this smallest of rabbits, it may be found in great numbers. No matter when or where one of these may be seen, a clump of chapparal or its burrow seem always at hand; thus it does not travel far, and a few jumps bring it to a place of safety. ... The burrows usually run into sand hillocks formed around bushes; sometimes, however, they are dug into the bare compact surface." (J. H. Clark, in Spencer F. Baird, 'Mammals of N. America.—Part ii. Special Report upon the Mammals of the Mexican Boundary,' p. 48, 1859.)

‖ *Cf.* also Bachman on *Sylvilagus palustris*: "Instead of leaping like the common Hare, it runs low to the ground, darting through the marsh in the manner of the Rat." (J. Bachman, "Descr. of a new Species of Hare found in South Carolina," Journ. Acad. Nat. Sci. Philad. vii. p. 196. Read May 10th, 1836.)

the Wood-Rabbit, though it certainly appears to get over the ground quite cleverly, particularly to one who has just missed, by under-shooting, a running shot " *.

The most remarkable member of the family, as to its habits, is the " *Romerolagus Nelsoni*, Merr.," from Mount Popocatepetl, Mexico, of which it is stated:—" This singular animal has exceedingly short hind legs, and instead of moving by a series of leaps, like ordinary rabbits, runs along on all fours, and lives in runways in the grass like the meadow-mice " †. Mr. E. W. Nelson, the discoverer of this creature, has furnished the following further particulars:—" A search under the overhanging masses of long grass-blades showed a perfect network of large arvicola-like runways tunneling through the bases of the tussocks, and passing from one to another under the shelter of the outcurving masses of leaves. It was evident that the rabbits were very numerous here . . . So far as observed, these animals are strictly limited to the heavy growth of saccatan grass, between about 3050 and 3650 meters . . . They make their forms within the matted bases of the huge grass tussocks, by tunneling passage-ways along the surface of the ground through the mass of old grass leaves and stems, and then hollowing out snug retreats within the weather-proof shelter thus obtained " ‡.

I am unfortunately unacquainted with the limb-skeleton of this interesting animal. Although from the foregoing description it results that it cannot be considered a burrowing animal, I venture to anticipate that its ulna will be found at least as little reduced as in the common Rabbit, and not placed behind the radius.

Hodgson § gives the following information on the habits of *Caprolagus hispidus* (Pears.):—" The Hispid Hare is a habitual burrower, like the Rabbit ; but, unlike that species, it is not gregarious, and affects deep cover, the pair dwelling together, but apart from their fellows, in subterranean abodes of their own excavation . . . Less highly endowed with the senses of seeing and hearing than the Common Hare or Rabbit, and gifted with speed far inferior to that of the former or even of the latter species, the Hispid Hare is dependent for safety upon the double concealment afforded by the heavy undergrowth of the forest ‖ and by its own burrow, and accordingly it never quits the former shelter, and seldom wanders far from the latter, whilst the harsh hair of its coat affords it an appropriate and unique protection against continual necessary contact with the huge and serrated grasses, reeds, and shrubs in the midst of which it dwells, and

* Elliott Coues, " Observations on the Marsh-Hare," Proc. Boston Soc. of Natural History, xiii. pp. 87, 88, 89 (1869).

† C. Hart Merriam, " *Romerolagus Nelsoni*, a new Genus and Species of Rabbit from Mount Popocatepetl, Mexico," Proc. Biol. Soc. Washington, x. p. 169 (1896).

‡ *Op. cit.* pp. 169, 170.

§ B. H. Hodgson, " On the Hispid Hare of the Saul Forest." J. A. S. Bengal, xvi. 1, pp. 573, 574 (1847).

‖ By later writers it is denied that *C. hispidus* is an inhabitant of the forest. Blanford (' Fauna of British India,' Mammalia, ii. p. 454, 1891) says :—" According to Hodgson the Hispid Hare inhabits the *Sál* forest, whilst Jerdon states with more probability that it is found in the Terai (that is, of course, the marshy tract usually thus called), frequenting long grass and bamboos &c." Jerdon's words are :—" It frequents jungly places, long grass, and bamboos, and, from its retired habits, is very difficult to observe and obtain " (T. C. Jerdon, ' Mammals of India,' p. 226, 1867).

dwells so securely that it is seldom or never seen even by the natives, save for a short period after the great annual clearance of the Tarai by fire ; and they tell me that it feeds chiefly on roots and the bark of trees, a circumstance as remarkably in harmony with the extraordinary rodent power of its structure as are its small eyes and ears, weighty body, and short strong legs, with what has been just stated relative to the rest of its habits. The whole forms a beautiful instance of adaptation without the slightest change of organism " *. Even if it had not been expressly stated, I would have concluded from the structure of the fore limbs (text-figs. XXXVI–XL) that the Hispid Hare is a burrowing animal : in fact, the only member of the family whose organization betrays fossorial propensities.

Nothing is known about the habits of the Sumatra Hare, *Caprolagus* (*Nesolagus*) *Netscheri* †. From the structure of its fore limbs, Pl. **38**. fig. 28, it may be safely inferred that it is a bad runner, and it may be an occasional burrower ; but it is certainly much less fossorial than *C. hispidus*.

The mode of locomotion of *Lagomys* (*L. pusillus*) is thus described by Pallas :— " Incedunt *L. pusilli* clumbi et subsultante gressu, sed propter brevitatem pedum, maxime posticorum, neque celeriter currunt, nec nisi inepte exsiliunt. In posticos pedes raro eriguntur " ‡ Winge concludes from this that " the mode of locomotion is therefore the same as in *Lepus*." Besides, he thinks it probable that the ancestors of *Lagomys* have been better runners than the recent species ; this, on account of the resemblance of the rump- and limb-skeleton between *Lagomys* and *Lepus*. Also, according to the same writer, some features in the skull of *Lagomys*, showing that the organs of smell and sight are less developed, point nevertheless towards a former different condition §. As seen from the figures (Pl. **38**. fig. 20), *Lagomys* resembles ordinary Rodents and Insectivores in the lateral position and non-reduction of the ulna, and also in its comparatively short hind legs. This is the primitive condition. Are we, then, to assume that the ancestors of *Lagomys*, starting from this condition, reduced their ulna and shortened their hind legs, only to revert again to the former primitive condition presented by the living species? Equally far-fetched seems to me the supposition that the choanæ had formerly been wider and the eyes larger. Neither *Prolagus* (Pl. **39**. fig. 36) nor *Titanomys* supports the former assumption, and there is no indication of larger orbits in *Prolagus*, nor of supraorbital processes in either of the two fossil genera. The statement, " incedunt *L. pusilli* clumbi et subsultante gressu," which recalls Cones's description of *S. palustris* (" scuttles along with shorter, quicker steps, more constrained and spasmodic, moving by jerks, as it were "), proves, in my opinion, an *incipient* stage of the leporine locomotion.

---

* The view expressed in the latter part of the last sentence is not correct.

† H. Schlegel, " On an anomalous Species of Hare discovered in the Isle of Sumatra : *Lepus Netscheri* " (' Notes from the Leyden Museum,' vol. ii. note xii. p. 59, 1880).

‡ ' Novæ Species Quadrup. e Glirium Ordine,' p. 35 (1778).

§ H. Winge, ' Jordfundne og nulevende Gnavere (Rodentia),' p. 113.

FIFTH CARPAL RAY.

The Pisiform.

Krause describes the pisiform of the domestic Rabbit as articulating with the ulnare on its volar side [*]; in the description of the ulna [†], it is stated that the distal termination of this bone has a condyle for the facet of the ulnare. These two statements imply that the pisiform of the domestic Rabbit articulates—as in Man— with the ulnare only. If they are correct, the German domestic Rabbits are different from those of this country; for in the English domestic and wild Rabbits I find the bone called pisiform articulating with the ulna as well as with the ulnare; this is the case moreover in all Leporidæ (Pl. 38. fig. 2, text-figures XXXI, XXXIV, XXXVIII), in all Lagomyidæ (Pl. 38. fig. 4), and in the great majority of Mammalia. In the Leporidæ the pisiform, the proximal part of which extends considerably in a transverse direction on the volar side of the carpus, shows even two facets for the volar side of the ulna.

From the following statements it appears that the so-called pisiform of Mammalia is a compound bone.

Daubenton mentions three accessory bones in the carpus of *Hylobates* and *Innus ecaudatus*: one of them is, in *Hylobates*, situated as follows : " il se trouve placé sur le joint qui est entre le troisième et le quatrième os du premier rang ; " situated, therefore, on the articulation between the ulnare and pisiforme [‡]. The carpal bones of *Innus* are said to have the same position as in *Hylobates*, only differing in their form [§]. In *Papio* the accessory bone in question is said to be wanting [‖].

Cuvier's description is almost identical. Speaking of the " ossified nodules in the muscle tendons " of the carpus, he says :—" Il y en a deux par exemple, dans le *gibbon* et le *magot* : l'un dans le tendon du cubital externe, sur le joint du pisiforme avec le cunéiforme . . . . manque dans les *sapajous* " [¶].

Leboucq describes and figures [**] a case in the Gibbon :—" Chez un Gibbon (*Hylobates leuciscus*) de la collection de l'Université de Gand, il existe entre le cubitus et le cubital du carpe un nodule osseux articulé latéralement avec le pisiforme (*p'*. fig. 28). Ce nodule me semble représenter le cartilage qui disparaît chez l'homme." (Reference is here made to the previous description of a cartilaginous nodule which is constantly met with in human embryos of the third and fourth month.) " En même temps que le crochet terminal du cubitus s'accuse nettement, il se développe dans le ménisque embryonnaire un nodule cartilagineux elliptique, faisant suite d'une part à la pointe du crochet et de l'autre se dirigeant vers l'extrémité proximale de l'intermédiaire." It disappears

---

[*] W. Krause. ' Die Anatomie des Kaninchens in topogr. und operativer Rücksicht,' 2te Auflage, p. 120 (1884).

[†] *L. c.* p. 119.

[‡] Buffon et Daubenton. Hist. nat. gén. et partic. xiv. p. 105 (1766).

[§] *L. c.* p. 127.   [‖] *L. c.* p. 151.

[¶] Leçons d'Anat. Comp. 2ᵉ éd. i. p. 425 (1835).

[**] H. Leboucq, " Rech. sur la Morphologie du Carpe chez les Mammifères," Arch. de Biologie, publ. par Van Beneden et Van Bambeke, v. p. 83, pl. iv. fig. 28 (1884).

constantly after the fourth month *. Leboucq considers this cartilaginous nodule of the human fœtus the homologue of the ossicle in *Hylobates*; both are parts of the pisiform, the pisiform of human anatomy being, in his opinion, but the distal epiphysis of the complete pisiform †. In a later paper the cartilaginous nodule is homologized with the os trigonum (tarsi) : "je crois donc pouvoir considérer ce nodule et l'os trigonum comme homologues " ‡, whence it would follow that the ossicle of *Hylobates* is equally the homologue of the trigonum.

The ephemeral cartilage of the human embryo has since been discovered in an ossified condition in a carpus of an adult, and received the name triquetrum secundarium §. Both this cartilage in the fœtus and the triquetrum secundarium occupy a more radial position than the ossicle of the Gibbon, wherefore it would appear that they are not, after all, the homologues of the latter, and this is proved to be the case by the discovery by Kohlbrügge of *two* accessory ossicles in the Gibbon. In three specimens of the three species *Hylobates leuciscus, H. agilis,* and *H. Mülleri,* an ossicle is situated between the styloid process of the ulna, the pisiform, and the ulnare. It rests on the processus styloideus and articulates with it and the ulnare. The pisiform joins the carpus at the point of junction between the ossicle and the ulnare. Kohlbrügge recalls the description of Daubenton, in whose honour the ossicle is named (*ossiculum Daubentonii*); and he adds that Camper had seen it in the *Inuus* ‖. In the carpus of a *Hylobates syndactylus* the following condition is described :—" Situated between the radius and the ulnare is an ossicle, which is joined to the radius and to the ossiculum Daubentonii by a fibrous ligament ; between both is cartilaginous tissue." The ossicle which, to all appearance, is that described by Camper in the Mandrill —and which has hence received the name *ossiculum Camperii*—was present in both hands of the Gibbon ; in the left manus the ossiculum Daubentonii was reduced to a small osseous nucleus ¶. From its position, the ossiculum Camperii corresponds to the cartilaginous nodule discovered by Leboucq in the human fœtus, and is therefore the homologue of the triquetrum secundarium (triangulare) of Man **. There can be no doubt that the ossiculum Daubentonii is the element which Leboucq has described in an adult *H. leuciscus*, since they occupy the same position. In Leboucq's figure—dorsal aspect of the carpus—the pisiform (*p.*) has been removed backward, in order to bring it into evidence ††.

---

* *Op. cit.* p. 81, pl. iii. fig. 17.   † *Op. cit.* p. 83.

‡ H. Leboucq. " Sur la Morphologie du carpe et du tarse," Anat. Anz. i. p. 20 (1886).

§ Pfitzner, " Bemerkungen zum Aufbau des menschl. Carpus," Verh. Anat. Ges. 7. Vers. in Göttingen 1893 (Ergänzungsheft Anat. Anz. viii. p. 191 (1893).—See also Morph. Arb. iv. p. 508 (1895).

‖ J. H. F. Kohlbrügge, " Versuch einer Anatomie d. Genus *Hylobates*" (M. Weber, Zool. Ergebn. einer Reise in Niederländisch Ost-Indien. i. pp. 338, 339, pl. xvii. fig. 9 (1890-91).

¶ *Op. cit.* p. 339, pl. xvii. fig. 10.

** The ossiculum Camperii (triquetrum secundarium, triangulare) or, as Thilenius terms it, os intermedium antebrachii, has been found in *Homo, Hylobates,* and *Inuus*, as mentioned in the text, and, by Pfitzner, in a carpus of *Phascolomys*. Pfitzner's specimen is figured and described by Thilenius (Morph. Arb. v. p. 9, pl. i. fig. 12 (1895)). I find what I take to be the same bone in Lemurs, Insectivora, and Rodentia, whereon more will be said in another place. (See P. Z. S. London. 1899. pp. 428–437.)

†† *Op. cit.* p. 191 (explan. of fig. 28).

Leboucq's view that the human pisiform is the homologue of the mammalian pisiform *minus* the ossicle he figures in the Gibbon receives confirmation by a discovery of Pfitzner's in the human adult carpus. He found in five cases a proximal process of the pisiform *. To this "*pisiforme secundarium*" would correspond the "*ulnare antebrachii*" of Thilenius, met with in ten manus of five embryos, where it is situated volad and ulnad of the proc. styl. ulnae, and proximally from the pisiform †. Both German authors take this element to belong to the same category as the os Camperii, viz. to be a carpal of a "preproximal series." We have, however, seen that Leboucq shows that the os Daubentonii, which in *Hylobates* is not unfrequently an independent ossicle, is contained in the mammalian pisiform. For my part, I see no stringent reason to assign this os Daubentonii to a "preproximal" series; from its position I consider it to be the first, proximal, carpale of the fifth ray, and it might therefore appropriately be designated as V. 1; it corresponds to the I. 1 on the radial side, the radiale marginale, which in *Echidna* actually articulates with the radius (Owen). In Reptilia, especially in Emydidae, we frequently find an ossicle or a cartilage occupying the position of a V. 1. Its absence in the Urodela is easily explained by the reduction of the ulnar part of the urodele carpus, even the fifth digit being lost. The reduction of the ulna and the ulnad extension of the ulnare may account for its being, in Mammalia, generally situated on the volar face.

What, then, is the distal part of the mammalian pisiform? One might suggest, as the easiest expedient for getting rid of this embarrassing element, that it is V. 2, viz. the second carpal of the fifth ray.

But, besides there being, as we shall see hereafter, another competitor for this distinction, there is not the slightest evidence of the distal pisiform having at any time occupied a similar position. On the other hand, it shows evidence of a former greater complexity. In most, if not in all Mammalia, except Man and the Anthropoids, the pisiform is provided with a distal epiphysis; and in some there is more than that. In the Rodent *Bathyergus maritimus*, as described by Von Bardeleben, ". . . the praepollex and the postminimus are both very well developed. The latter consists of two bones, of which the proximal (*pi p.*) is the true pisiform, and measures 5 millim. in length, while the distal is 7·5 millim. in length. We must therefore in the future distinguish a proximal from a distal 'pisiform,' and I regard the former as, in all probability, the carpal, and the latter as the metacarpal segment of the postminimus" ‡.

Two skeletons of *Bathyergus maritimus* are in the Natural History Museum, neither of them quite adult. In the older one, which is the original of Von Bardeleben's figure 3, the distal part of the pisiform is incompletely ossified, as shown in the figure; it is still completely cartilaginous in the younger specimen. A similar, more or less ossified distal

---

* Morph. Arb. iv. p. 508 (1895). "Dieser Fortsatz war (in vier Fällen) proximal, und zugleich eher etwas dorsal als volar gerichtet. Seine plane Fläche stellt eine continuierliche Fortsetzung der Gelenkfläche des Hauptstückes dar; im Übrigen war der Fortsatz ringsherum durch eine tiefe Einziehung abgesetzt."

† Morph. Arb. v. p. 470 (1896).

‡ "On the Praepollex and Prachallux, with observations on the Carpus of *Theriodesmus phylarchus*," Proc. Zool. Soc. London, 1889, p. 260, pl. xxx. fig. 3, *pi.p.*, *pi.d.*; id. Verh. Anat. Ges. 3te Vers. Berlin (Ergänzungsheft) Anat. Anz. iv. p. 108 (1889).

pisiform I find in the hystricine *Ctenomys* and in *Mus*, and it will probably be met with in many other fossorial and climbing Rodents.

What seems to be a remarkable adaptation of the distal pisiform to a special function is exhibited by the strong cartilage, which in *Pteromys* is prolonged to support the lateral membrane serving as a parachute. Thilenius makes of it an element of an antebrachial series, his " ulnare antebrachii " * ; but he is misled by Owen's much reduced figure of the skeleton of a " *Pteromys colucella* "†, in which the detached cartilage has been drawn proximally to the pisiform and separated from it by a small interspace. The true connection of this cartilage was already known to Buffon ‡. He described it as a bone ; but in the only skeleton (*Pteromys magnificus*) at the Natural History Museum in which this element is preserved it is perfectly cartilaginous, and as such it is described by Owen in *Sciuropterus colucella* §. In *Pteromys magnificus* it is chiefly attached to the *distal* end of the pisiform, and, by a much smaller process, to the tuberosity of the fifth metacarpal. Its direction is in the beginning right backward, in the prolongation of the long axis of the osseous pisiform ; but gradually it turns upward, forming in its entirety a semicircle. It might be maintained that the patagial cartilage of Sciuropterini is in origin quite extraneous to the pisiform, and that it has only secondarily become supported by this widely projecting bone. With the scanty material at my disposal, I am not in a position to follow up the matter closer, nor is this the place to do so. A clue might be obtained from young specimens of *Pteromys*; and if they should show both the usual pisiform epiphysis and the patagial cartilage, they would support the view of an extraneous origin of the latter.

The lengthened subcylindrical bone which in the insectivore *Chrysochloris* extends from the carpus to the humerus, " simulating a third antebrachial bone," was considered by Meckel ||, followed by Carus ¶, Peters **, Giebel ††, and Dobson ‡‡ as an ossification of a tendon ; regarded by the latter three as that of the m. flexor digitorum profundus. Cuvier §§, A. Wagner ¶||, Gervais ¶¶, and Owen *** homologize this bone with the pisiform.

---

* Morph. Arb. (Schwalbe) v. p. 508 (1895).

† ' Anatomy of Vertebrates,' ii. p. 385, fig. 154 (1866).

‡ " Il y a de plus dans le polatouche un os (AA) long de 5 lignes, en forme d'arête ou d'éperon, qui tient au quatrième os du premier rang du carpe, et qui s'étend obliquement en arrière et en haut le long du bord de la membrane qui forme les ailes de cet animal." (Hist. Nat. gén. et partic. x. p. 113, pl. xxiv. 1763.)

§ L. c.

|| System d. vergl. Anat. ii. (2) p. 374 (1825). He calls the element " ein vom Streckknorren des Oberarmbeins zum Speichenende [it is, however, on the ulnar side] der Handwurzel gehendes, starkes, verknöchertes Band."

¶ " . . . ein dritter Knochen des Untergliedes, welcher jedoch nur als eine verknöcherte Sehne, oder vielmehr ganz verknöcherter Muskel (*flexor carpi ulnaris*), anzusehen ist."—C. G. Carus, Erläuterungstafeln zur vergleichenden Anatomie, ii. p. 31, Taf. 9, fig. 19, *b*" (1827).

** W. Peters, Naturw. Reise nach Mossambique, Zoologie, i. p. 72 (1852).

†† Giebel, in Bronn's Klassen u. Ordnungen, vi., v. p. 534 (1879).

‡‡ G. E. Dobson, 'A Monograph of the Insectivora,' p. 121 (1882).

§§ G. Cuvier, Leçons d'Anat. Comp. 2ᵉ éd. i. p. 426 (1835). Schreber's ' Säugthiere,' Suppl. ii. p. 120 (1811).

¶¶ P. Gervais, Hist. Nat. des Mammifères, i. p. 252 (1854).

*** R. Owen, ' On the Anatomy of Vertebrates,' i. p. 302 (1866).

From the description given by Dobson, it becomes quite evident that from the distal end of this bone there arise tendons for the four digits, so that we have here a bone functioning as the common tendon of the flexor digit. prof. From this, however, it does not necessarily follow that it is an ossified tendon. (The pisiform of Man is imbedded in the tendon of the m. ulnaris internus : but scarcely any anatomist will to-day persist in considering it to be a tenontogenous sesamoid. It has been degraded to play the part of a "sesamoid" *, and that only in Man and some of the Anthropomorpha.)

Dobson has figured the volar aspect of the carpus of a *Chrysochloris Trevelyani* †, in which the alleged ossification of the m. flexor prof. tendon has been removed. Here we see, ulnad from the lunar, the flattened face of a bone (*us.*), which is not referred to in the text; in the explanation of pl. xiii. fig. 5 it is termed the "ulnar sesamoid." Carus ‡ has seen and described this ossicle, and so have D'Alton sen. & jun. § The first-named states that the "ossified tendon" starts (" ausgeht ") from it ; both Carus and the D'Altons call it a pisiform (" Erbsenbein"); but, so far as I am aware, later authors, with the exception of Dobson, have overlooked it.

In a skeleton of *Chrysochloris aurea*, this so-called sesamoid articulates dorsad with the ulnare, dorsad and radiad with the lunar, proximally with the ulna, volad and distally with the "flexor dig. prof. ossification." The latter shows at the dorsal side of its distal base two facets, the larger ulnad one for the " ulnar sesamoid," the smaller radiad for a volar and distal projection of the lunar.

I take this " ulnar sesamoid " to be the ossiculum Daubentonii, viz. the basal part of the pisiform ; but, owing to the distorted condition of the *Chrysochloris* carpus—the lunar articulates with both radius and ulna—and from my insufficient material, which consists in a single skeleton of one of the smallest species, I cannot state my case with greater certainty. If my view is correct, then the "tendon ossification" is in all likelihood the homologue of the distal part of the pisiform of other Mammalia, where it very often starts backward at right angles from the long axis of the limb, sometimes, as in *Hylobates* ‖, directly downward, and sometimes more or less upward, viz. in a proximal direction (*Talpa*). Which is the primitive direction I cannot for the present decide. The Chrysochloridæ vary so much from one species to the other that Cope has divided them into three genera ¶ ; and we may hope that it will be possible to settle the question of the homology of this curious bone when the skeletons of these different forms shall have become available for comparison.

It appears to me that the distal part of the pisiform will prove to be a remnant of a lateral ray, which only secondarily entered into connection with the ulnare and the ulna. Of this lateral ray the other accessory distal elements of *Bathyergus*, *Ctenomys*, *Mus*, and

---

* " Das Pisiforme spielt . . . die Rolle eines in der Sehne des Muskels (*flexor carpi ulnaris*) befindlichen Sesambeins." Gegenbaur, Lehrbuch d. Anatomie des Menschen, 6te Aufl. i. p. 422 (1895).

† ' A Monograph of the Insectivora,' pl. xiii. fig. 5 (1882).   ‡ *L. c.*

§ E. D'Alton d. Ae. und E. D'Alton d. J., ' Die Skelete der Chiropteren und Insectivoren,' p. 22 (1831). Kohlbrugge, *l. c.* fig. 10.

¶ ' American Naturalist,' xxvi. p. 126, footnote 1 (1892).

even the cartilage of Sciuropterini, possibly were parts. There is not the slightest evidence that the lateral ray has ever been a digit of the manus of the Tetrapoda.

Carpale 5 (V. 3).

The question whether there is some ground for assuming a central carpale (V. 2) in the fifth ray is closely connected with the present subject, so that it will be dealt with in this place.

I have known for a long time a comparatively large facet on the proximal ulnar side of Metac. V in two species of the fossil *Prolagus*, *P. oeningensis* (Kön.), and *P. sardus* (Wagn.) (Pl. 38, fig. 19, *r*), for which I could not account, the metacarpals of *Lepus*, which were at my disposal at the time, showing nothing of the kind. This same facet I have of late found to be present in *Lagomys* (*L. rufescens*), where it articulates with a small ossicle, which also presents a facet to the ulnare (Pl. 38, fig. 4, *r*). The ossicle is likewise present and has the same connections in *Sylvilagus brasiliensis* (text-figures XXIX & XXXI), *S.* sp. from Bogotá (Pl. 33, figs. 1, 2, *r*), and *Oryctolagus crassicaudatus* (text-figures XXXII & XXXIV). In two other species (*Nesolagus Netscheri* and *Caprolagus hispidus*) the facets are visible, but the ossicle has been lost.

What is the ossicle in question?

As is well known, Gegenbaur was the first to express the opinion that the mammalian hamatum is a compound of carpalia 4 and 5, on the ground that in lower forms we find the fourth and fifth digits provided each with a separate carpale [*]. Leboucq sees in the mammalian hamatum the homologue of carpale 4 only. "Le carpien 4+5 de Gegenbaur ne correspond exactement dans les premiers stades de développement qu'au métacarpien IV seul; le métacarpien V est placé latéralement par rapport à ce carpien. Le carpien 4+5 se sépare de l'axe au niveau de l'intermédiaire; quant au Vᵉ métacarpien, tout fait supposer que son rapport avec le dernier os de la rangée distale est secondaire chez les mammifères; primitivement c'est toujours avec le IVᵉ métacarpien seul que ce carpien est en continuité. On ne voit à aucun stade de développement ce carpien formé de deux parties, on présentant le moindre vestige de sa double origine. Où serait alors le carpien 5? En examinant les premiers stades de développement, non-seulement chez l'homme, mais chez les divers mammifères que j'ai pu étudier, on voit que le métacarpien V est placé en face de l'os cubital, mais séparé de lui par un interstice plus grand que celui qui sépare les autres métacarpiens de leur carpien correspondant. On peut admettre que c'est au niveau de cet espace que doit se trouver le carpien 5. Quant à déterminer ce qui doit représenter ce carpien, on peut admettre son absence complète, ou bien le considérer comme non différencié, et contenu virtuellement dans un des éléments squelettiques du voisinage : soit l'os cubital, soit le métacarpien V. L'hypothèse la plus probable serait de considérer le carpien 5 comme ne s'étant pas différencié à l'extrémité proximale du métacarpien V " †.

---

[*] Untersuchungen z. vergl. Anatomie d. Wirbelthiere, i. pp. 45, 53, 121 (1864).

† "Rech. sur la Morphologie du Carpe chez les Mammifères," Arch. de Biologie, publ. par E. van Beneden et Ch. van Bambeke, pp. 92, 93 (1884).

In the following year Turner described and figured five distinct distal carpal bones in a Whale. After having mentioned that in *Mesoplodon bidens* "carpalia 4+5 formed a single bone . . . . which was grooved on its dorsal surface opposite the interval between metacarpals IV and V," he proceeds to describe the carpus of an adult *Hyperoodon rostratus.* "The distal carpalia are five distinct bones, not so regularly faceted as those in the proximal row, and with a larger proportion of cartilage between them. These bones pass from the radial to the ulnar border in regular order, as C 1 to C 5, and each is associated with the metacarpal bone of its corresponding digit. A similar arrangement exists in both limbs, and the carpus possesses also an elongated pisiform cartilage, which in one is partially ossified " *.

Von Bardeleben had previously made the following statement :—" Deutliche Anzeichen einer früheren Trennung in zwei Elemente zeigt das Hamatum bei den Beutelthieren, weniger auffallend bei den Nagern, sowie bei *Ziphius* (*Hyperoodon*). In zwei Stücke getrennt, aber, auf der einen Seite wenigstens, schon im Verwachsen begriffen, ist es an dem Skelete eines jungen Bären in Berlin." †.

To these assertions Baur replied that he had never in any mammalian embryo observed the hamatum to be the outcome of a fusion of two elements, and he adds :—" Wenn es bei älteren Thieren den Anschein hat, als wäre eine Theilung vorhanden, so ist dies eben etwas secundäres und ist morphologisch nicht verwendbar " ‡. In his latest utterances on the subject §, Von Bardeleben mentions only the separation of the hamatum in " *Ziphius* (*Hyperoodon*)," thus tacitly withdrawing the statements regarding other Mammalia, made at the meeting of the Jenaische Gesellschaft of May 15, 1885, above quoted, as well as in a subsequent meeting of October 30 ‖.

The manus of the Jena specimen of *Ziphius cavirostris*, to which Von Bardeleben refers, has been described and figured by Kükenthal. It contains altogether three distal carpalia : the one resting on Metac. IV and V shows on its dorsal surface a delicate furrow, " eine zarte Furche als Andeutung einer früheren Trennung zweier Carpalia " ¶. This is what Von Bardeleben, in his " Referat," calls having found in *Ziphius* "eine natürliche Zerlegung des ' Hamatum ' in das Carpale IV und Carpale V " **, and further on : " Dass Ref. im Mai 1885 die primitive (vielleicht secundäre—jedenfalls dem Verhalten bei Urodelen entsprechende) Trennung des ' Hamatum ' oder Carpale 4+5 (Gegenbaur) in Carpale 4 und Carpale 5 bei *Ziphius cavirostris* auffand (an der Hand des Jenaer Exemplars)." ††.

In his subsequently-published researches, Kükenthal describes fresh facts and sums up those previously recorded ‡‡. In embryos of *Beluga* and *Monodon* there sometimes

---

* Journ. Anat. Physiol. xii. pp. 180, 183 (1886).

† Jenaische Zeitschr. f. Naturw. xix. (xii.), Suppl. ii. p. 87. Sitzung am 15. Mai 1885.

‡ Zool. Anz. 1885, p. 487.

§ Proc. Zool. Soc. London, 1894, p. 375 ; " Hand und Fuss," Verh. d. Anat. Ges. viii. pp. 263, 301 (1894). Jen. Zeitschr. xix. (xii.), Suppl. iii. Sep.-Abdr. p. 78 (1885).

¶ Denkschr. d. med.-naturw. Ges. zu Jena, iii. pp. 38, 46, pl. iii. fig. 18 (1889). See also E. Rosenberg, *op. cit.* p. 2, footnote 4 : Kükenthal, in Morph. Jahrb. xiv. p. 56 (1888).

** *Op. cit.* p. 263.   †† *Op. cit.* p. 301.

‡‡ Denkschr. med.-naturw. Ges. Jena, iii. pp. 268–280 (1893).

occur five carpalia. An additional instance of five carpalia in an adult *Hyperoodon* is adduced from a specimen in the Royal College of Surgeons [*], and two examples in embryos of the same genus [†]. The reduction in the number of carpalia is explained by fusion or vanishing ("Schwund"); the fusion is brought about in two different ways:— "Bei den *Ziphioiden* verschmilzt das Carpale distale 5 mit dem C. dist. 4, es kommt also zur Bildung eines Hamatums; bei den *Delphiniden* verschmilzt das Carpale distale 5 mit dem Ulnare, oder aber es kommt überhaupt zu einem völligen Schwunde des ersteren, und seine Stelle wird vom Ulnare eingenommen." Transitions between both types of reduction occur in *Beluga* and *Monodon*.

In an embryo of *Emys lutaria*, of 8 mm. length, Rosenberg found in the place of one hamatum two completely-separated cartilages. "Der mehr ulnar gelegene ist etwas kleiner und steht ausser mit dem Ulnare und seinem radialwärts gelegenen Nachbarelement nur mit dem Metc. IV in Beziehung. Der andere der in Rede stehenden Knorpel trägt das Metc. IV; in seinem dorsalen Abschnitt wird er auch von dem Metc. III berührt, welchem übrigens sein eigenes Carpale zukommt. Es ist kein Zweifel, dass diese beiden ovoiden Knorpel die zu postulirenden Carpale 4 und Carpale 5 sind, die in typischem Verhalten zu ihren Metacarpalien vorliegen . . . es stellen daher das Carpale 4 und Carpale 5 in diesem Stadium volkommen selbständige Elemente dar." In three larger embryos (10 mm.) the same investigator observed three stages of fusion of the carpalia in question. He considers that this result supports Gegenbaur's view with regard to the hamatum of Mammalia [‡].

Pfitzner has given the name *Os Vesalianum* to an ossicle in the human carpus, first described by Vesalius, who considered it to be a sesamoid. It is situated on the ulnar side of the hamatum, and its distal facet touches the tuberosity of the fifth metacarpal [§]. Later on, he mentions two other cases in Man, one found by Gruber [‖] and a third by himself [¶]. In Vesalius' case, the ossicle articulated apparently with the hamatum and Metac. V. In Gruber's case "begann es vom Hamatum abzuwandern und sich dem Met. V enger anzuschliessen, mit dem es wahrscheinlich schon coalescierte." In Pfitzner's own case finally, the ossicle had no more connection with the hamatum, and had undergone a synostosis with the Metac. V. Pfitzner continues: "Als weitere Rückbildungsstufen haben wir wohl anzunehmen, dass es vom Met. V gänzlich assimilirt wird und in dessen Tuberositas aufgeht," a view which is confirmed by what Thilenius, who terms this element "Carpo-metacarpale 8," has found in the human embryo [**]. Pfitzner is of opinion " dass in gewissem Sinne das Os Vesalianum, namentlich in seiner ursprünglichen Lage, einem hypothetischen Carpale V zu entsprechen vermöchte." Like their predecessors, neither Pfitzner nor Thilenius have met in the human carpus with a division of the hamatum into a carpale 4 and 5, in Gegenbaur's sense.

Pfitzner's os vesalianum carpi occupies about the same position as the ossicle in

[*] *Ibid.* p. 278, text-fig. 11.   [†] *Ibid.* text-figs. 12, 13.
[‡] Morph. Jahrb. xviii. pp. 8, 9 (1892).
[§] Morph. Arb. i. p. 756 (1892).
[‖] Arch. f. Anat. Phys. pp. 499, 500, Taf. xii. (1870).
[¶] Morph. Arb. iv. pp. 543, 544 (1895).   [**] *Ib.* v. pp. 488, 489 (1896).

Lagomyidæ and Leporidæ mentioned above. I am not aware that it has ever been recorded before in lagomorphous Rodentia; while it seems quite a common element in Mammalia provided with a well-developed fifth digit, at any rate in Rodentia, Insectivora, and Edentata, and was known to the older anatomists. Cuvier mentions it in the Great Armadillo (*Priodon giganteus*), and describes and figures it as situated laterad of the ulnare and articulating with the Metac. V *. In the figure published by Flower †, it would appear to articulate with the ulnare as well. As to its presence in Rodentia, Cuvier remarks: " Enfin il y a très souvent aussi au bord externe du carpe, en dehors du cunéiforme et de l'unciforme, un os surnuméraire, petit et lenti-culaire ; on le voit dans le *castor*, le *porc-épic*" ‡. It is figured in a carpus of the *Castor* §. In the ' Leçons d'Anatomie comparée,' mention is again made of this "os surnuméraire" in the *Hystrix*: ". . . il y a un os surnuméraire entre le pisiforme et l'os métacarpien du cinquième doigt ; il est attaché sur l'os crochu" ‖.

Thilenius ¶, quoting Cuvier's figure of the *Castor* carpus, is inclined to consider this ossicle as his (Thilenius's) " ulnare externum "=the ulnar part of Pfitzner's triquetrum bipartitum of the adult, found in the human embryo **. He adds, however : " Infolge der radialen Verschiebung des Carpale (4+5) erreicht es indessen auch das Metac. V." The question is whether, when an os vesalianum is present, the hamatum is really displaced, or is not rather in its original position ; only secondarily either supplanting the vesalianum, or acting in a compensatory manner for it, when the latter is either displaced or has disappeared. When comparing Thilenius's figures 11 and 12 of this " ulnare externum " †† with figures 13 and 14 ‡‡, representing a later stage, the impression is conveyed that in the latter this ulnare externum (*ue*) has been displaced proximally by the ulnad extension of the hamatum. A secondary *proximal* displacement of a carpal (or tarsal) would, however, be quite unusual, and Thilenius has expressed some doubt §§ whether the figures mentioned all represent the same bone. In fig. 11, where *ue* abuts upon Metac. V, the former element might be Pfitzner's vesalianum (carpo-metacarpale 8, Thilenius). The text-figure XXXIV of the present paper seems to exclude the possibility, ventilated by Thilenius ‖‖, that " vesalianum " and " ulnare externum "— which have not yet been found together in the same manus of Man—might represent one and the same bone. The enormous ulnad and volad expansion of the ulnare, as shown for the Lagomorpha in this figure (XXXIV)—which occurs in other Mammals also—leads to the assumption of its being a compound of an ulnare+ulnare externum Thilen. The " ulnare externum " (=ulnar part of triquetrum bipartitum Pfitzn.) would then be the second (central) carpale of the fifth ray (V. 2).

Meckel has described the os vesalianum in *Erinaceus*:—" Der *Igel* hat in der obern, weit breitern Ordnung *vier* Knochen. Kahn- und Mondbein sind zwar verwachsen,

---

* Oss. foss. v. 1, p. 127 (1823).

† 'An Introduction to the Osteology of Mammalia,' 3rd ed. fig. 110 " *a*," p. 307 (1885).

‡ Oss. foss. v. 1, p. 48 (1823). § *Ib.* pl. ii. fig. 10.

‖ Leçons d'Anat. comp. 2de éd. i. p. 427 (1835).

¶ Morph. Arb. (Schwalbe), v. pp. 508, 509 (1896).

** Morph. Arb. v. pp. 473, 474 (1896).

†† Morph. Arb. v. pl. xxi (1896).   ‡‡ *Ib.*   §§ *Ib.* pp. 489, 508.   ‖ *Ib.* p. 489.

allein das grosse dreieckige Bein trägt aussen und vorn einen kleinen, runden Knochen eingelenkt, den man ein zweites Erbsenbein nennen kann. Von den vier vordern ist das Hakenbein weit kleiner als gewöhnlich, und das dreieckige stösst daher aussen beträchtlich weit an den fünften Mittelhandknochen " *.

Owen mentions the same ossicle in the Hedgehog, but more distally :—" A sesamoid is attached to the outside of the base of the metacarpal of the digitus minimus " †. In a left carpus of *Erinaceus europæus* lying before me, the ossicle articulates with both the ulnare and Metac. V, the facet for the latter being smaller and, as in *Priodon*, situated ulnad from the ulnare. The same bone is mentioned in *Gymnura* by Dobson ‡.

Referring to this ossicle, Leboucq says :—" Ce qu'on appelle 2° pisiforme, existant chez quelques mammifères (hérisson, tatou, etc.), n'est qu'un sésamoïde développé dans le tendon de l'extenseur cubital du carpe " §. It may be a matter of surprise that, in the same chapter in which Leboucq insists with strong arguments that the pisiform cannot be classed among " les os sésamoïdes," he casts aside with a few passing words this equally important bone. The explanation is to be found in the words " chez quelques mammifères ;" the author being evidently not sufficiently acquainted with the " os vesalianum."

Having placed the facts before the reader, I have now to sum up. All the attempts (Leboucq, Baur, Rosenberg, Pfitzner, Thilenius) to trace ontogenetically the presumed fusion of carpalia 4 and 5 to form the " hamatum " have confessedly failed. Gegenbaur explains this negative result by supposing that the Mammalia inherited the " hamatum," from lower Vertebrates. This leads him to the assumption that the occasional occurrence of two separate carpalia (4 and 5) among Cetacea is secondary ; the more so as we find other very considerable changes in the manus of these animals ‖.

To this argument might be opposed the daily increasing number of instances brought forward in which we see primitive characters occurring precisely in those species, or in those organs, which in other respects are highly differentiated (specialized), the preservation of old characters being obviously due to the specialization of others. This by no means new truth was, if I am not mistaken, first enunciated by Haeckel.

In support of the foregoing, I wish to refer to a very noteworthy remark by Gegenbaur himself. In defence of certain conclusions arrived at in his well-known " Gliedmaassenskelet der Enaliosaurier " ¶, he states that in Sauropterygia and Ichthyopterygia the

---

* System d. vergl. Anat. ii. 2, pp. 393, 394 (1825).   † 'Anatomy of Vertebrates,' ii. p. 390 (1866).

‡ 'A Monograph of the Insectivora,' p. 21 (1882).   § Arch. de Biologie, v. p. 84 (1884).

‖ " Die Einheitlichkeit des *Hamatum* der Säugethiere ist von mir als ein auf dem Wege der Phylogenese erworbener Befund erklärt worden, da in niederen Abtheilungen der vierte und fünfte Finger je ein discretes Carpalstück besitzen. Da jener Erwerb durch Concrescenz bald auf die Säugethiere überging, möchte ich bezweifeln, dass im Carpus der Cetaceen der niedere Zustand noch zu erweisen ist, selbst wenn auch unter den vielerlei dort bestehenden Befunden ein Carpale 4 und ein Carpale 5 sich darstellt. Denn die übrigen Veränderungen sind in diesem Handabschnitte zu bedeutend, als dass ein *secundär* erfolgtes Zustandekommen eines dem ursprünglichen ähnlichen Verhaltens zweier distaler Carpalia ausgeschlossen wäre." (C. Gegenbaur, Vergl. Anat. der Wirbelthiere, i. p. 542, 1898).

¶ Jen. Zeitschr. v. (1870).

adaptation to a new function does not in any way explain the typical features of their limbs. "*Where we meet with similar adaptations, the original condition has never been completely effaced*" (italics mine)*.

The undivided condition of the "hamatum" in terrestrial Mammalia can now be explained in a very simple and obvious manner, since by means of the "os vesalianum" we are enabled to show that the presence of a separate carpale 5 is not in the least limited to af ew cases among Cetacea, but is a frequent occurrence in other Mammalia likewise, a circumstance which has hitherto either been wrongly interpreted or entirely overlooked. The "hamatum" of Mammalia is not carpale 4+5 of Reptilia, but it is a carpale 4 which, as a rule, has become enlarged, and has, in addition to its own functions, usurped those of carpale 5. Whether a usurpation is *in every instance* to be assumed is another question, which cannot be entered into here; it may, for the present, be sufficient to repeat that the superadded function of carpale 4 may often be not the cause but the consequence of the degradation of carpale 5.

Where carpale 5 is absent in the terrestrial Mammalia, it has, so far as my experience goes, either disappeared by atrophy, or become absorbed by the tuberosity of Metac. V, as in Man. Finally, therefore, since the fusion of carpale 5 with carpale 4 has never been observed in these, its occurrence may be peculiar to the Cetacea.

Remarks on the Metatarsus and Tarsus of Lagomorphous Rodentia.

1. *Metatarsale* 1 *and Tarsale* 1.—Krause states † that in adult Rabbits the os tarsale 1 becomes fused with the os Metatarsi 1, and for this he refers to his text-figure 64 B. He continues as follows:—" In new-born animals, however (fig. 64 A), the tibial prominence of the proximal extremity of Metat. 1 is independent, and consists of an os tarsale and a lengthened distally-pointed bone, representing a rudiment of the hallux, at the distal end of which there is inserted the tendon of the m. tibialis anticus. In reality, therefore, the os tarsale 1 of the Rabbit is the o.t. 2 of Man, and the os Metatarsi 1 of the Rabbit represents the os tarsale 1, the hallux and os Metat. II of Man." So far as the fig. 64 A, "horizontal section of right hallux of a 12-days-old Rabbit," goes, this is correct, assuming that the two outline-figures of the tarsalia (1 and 2) are meant to show them in a cartilaginous condition. But the lettering of fig. 64 B, " right os Metat. 1 " (meaning Metat. II of comparative anatomists) of an adult

---

* ". . . muss daran festgehalten werden, dass die Anpassung an eine neue Function keineswegs das Typische der Gliedmaassenform zu erklaren vermag. Wo wir solchen Anpassungen begegnen, hat sich der ursprüngliche Zustand nie ganz verwischt. In der Flosse der Balaenen ist das Sängethierarmskelet klar zu erkennen, ebenso wie bei den Cheloniern die Schildkrotenextremität. Hier bei den Enaliosauriern ist auch gar nichts auf Reptilien Beziehbares am Flossenskelet vorhanden. Von der schon bei Amphibien vorhandenen Differenzirung von beiderlei Gliedmaassen nicht ein blasser Schein! Es müsste also an der Gliedmaasse ein Ruckgang bis zu den ersten Anfängen erfolgt und von diesen her eine selbständige Ausbildung eingetreten sein, wenn Beziehungen zum Reptilientypus hier einmal an der Gliedmaasse bestanden haben mögen. Jedenfalls gehören diese Bildungen nicht in die Reihe der Reptiliengliedmaassen, sondern unter die Anfänge, wie sie denn gerade in dem schon beregten Mangel des Different-werdens von Vorder- und Hinterextremität sogar unterhalb der bis jetzt bekannten Reptilien sich stellen. So birgt sich in diesen Fragen ein interessantes Problem." (Vergl. Anat. der Wirbelthiere, p. 531.)

† W. Krause, Anatomie d. Kaninchens, 2te Aufl., p. 132 (1884).

Rabbit, from the medial side, is erroneous. The process I, "place of the real Hallux," is the tuberosity of the Metat. II; with this tuberosity neither the tarsale 1 nor the rudiment of the Metat. I come in contact, and therefore they cannot form a connection with it. The proximal process of Metat. II, numbered 1 (=place [*Stelle*] of the real os tarsale primum), represents instead the rudimentary Metat. I (see Pl. 38. figs. 5 and 6 I), which in young *Lepus* is distinct, but afterwards becomes fused with Metat. II. Tarsale 1 is visible in the young Rabbit in a cartilaginous condition *, but in this species and in a *Sylvilagus* from Bogotá, in both of which I have been able to examine various stages, I have neither observed an ossification of it, nor a fusion with the rudimentary Metat. I, as assumed by Krause and by Leche. It gradually shrinks and apparently is absorbed †. It is quite possible that in some species a fusion may take place as a rule or exceptionally; but I deny it to have been demonstrated in the Rabbit, in which it is said to be the rule. Professor Howes informs me that he too has searched in vain for evidence of this.

2. *Fusion of Tarsale 2 with Metatarsale II.*—A fact hitherto not noticed in Lagomorpha is the fusion of tarsale 2, the mesocuneiform (c_2 of my figures) with Metat. II. This fusion takes place in *Prolagus* (Pl. 38. figs. 17, 27 *a*), in *Lagomys* (Pl. 38. figs. 16, 26 (c_2)), and in some Leporidae. In *Nesolagus Netscheri* (Pl. 38. fig. 23), the figured specimen of which is not adult, the fusion is not quite complete; in the older specimen at the Leyden Museum I saw it was complete. In a specimen of *Sylvilagus brasiliensis* from Lagoa Santa, the property of the Copenhagen Museum, tarsale 2 is fused in the right limb and distinct in the left; in an incomplete limb of the same species in the Royal College of Science, London, the fusion is complete.

3. *Praecuneiforme.*—As in the case of the vesalianum carpi (see pp. 501–3), my attention was arrested by an accessory bone in *Prolagus sardus* through a small facet on the tibial side of the proximal termination of Metat. II, or rather of Metat. I, since, as shown before, this part is occupied in the young by the rudimentary Metat. I, which later on becomes fused with Metat. II (Pl. 38. fig. 17. *pc*; fig. 27 *a*, facet on the upper left side of Metat. II). This is the region which corresponds to the insertion of the musc. tib. posticus, and therefore the ossicle, indicated by the facet, is the so-called distal prae-hallux, or Baur's "klauenartiges Gebilde." Winge has denied the existence of this ossicle in *Lepus* and *Lagomys* ‡, but I have found it in both families, and, as we shall see later, it has been met with as a rare occurrence even in *Lepus europaeus*. In *Lagomys* it articulates (Pl. 38. figs. 15, 16, 26 *pc*) by a smaller facet with the navicular as well, and lies in the distal continuation of a much larger ossicle (fig. 26, *ti*), which articulates with the navicular and the astragalus. The latter is undoubtedly Baur's and Leboucq's "tibiale" (the proximal ossicle of Von Bardeleben's "praehallux").

I find the smaller, distal, ossicle in the following Leporidae, viz. in *Nesolagus*

---

* See Leche, in Bronn's Klass. u. Ordn. d. Thierr. vi. v. 28ᵗᵉ Lief. pl. xcvi. fig. 3 (1885).

† Ketterer (Comp.-rend. et Mém. Soc. Biol. (10) i. p. 807, 1894) regards the ossicle, which I with others hold to be a rudimentary Metat. I, as tarsale 1, denying all trace of the former. The presence of a cartilaginous tarsale 1 in young Rabbits is easy of observation, but presumably it was not yet chondrified in the stages examined by Ketterer.

‡ H. Winge, "Jordfundne og nulevende Gnavere." E Museo Lundii, i. p. 169 (1887).

Netscheri (Pl. **38**. fig. 23, *pc*); in *Oryctolagus crassicaudatus* (fig. 22, *pc*), where it seems on its way to undergo a synostosis with Metat. II ; in *Caprolagus hispidus* (fig. 24, *pc*), where it has shifted its position completely to the volar side of Metat. II ; in a specimen of *Sylvilagus brasiliensis*, from the Copenhagen Museum; and lastly in the Wild Rabbit, where the ossicle is very small and situated volad as in *C. hispidus*. I owe this specimen to Mr. Sherrin, Articulator in the Nat. Hist. Mus., who at my request dissected some Rabbits' feet, in search of the ossicle in question.

In his careful researches " Ueber den Säugetier-Praehallux " [*], Tornier met with this ossicle in one case only of all the Rabbits' and Hares' feet examined, and great stress is laid on this isolated occurrence. "Die Lage dieses überzähligen Knöchelchens beweist unwiderleglich, dass es selbst homolog ist dem Knochen welcher bei vielen der bisher untersuchten Tiere der *I* 1-Medialseite gegenüber liegt. Da er an Hasenfüssen individuell auftritt und an jungen Kaninchen- und Hasenfüssen nicht vorhanden ist, so ist es zweifellos, dass er eine secundäre Bildung ist, und daraus ist mit Sicherheit zu schliessen, dass er auch bei den Tieren, wo er immer vorkommt, eine secundäre Bildung ist" [†], And again : " Der musc. hallucis abductor-Knochen kommt endlich drittens zuweilen bei erwachsenen Vertretern solcher Thierarten vor, bei welchen der Knochen unter normalen Umständen weder im Alter noch während der Ontogenese vorhanden ist (*Lepus timidus*) [‡] : bei diesen Individuen ist er—dagegen giebt es keinen Widerspruch— secundär entstanden " [§]. Therefore, as already stated in the first-quoted passage, he again asserts that the homologous bone in all other Mammals is equally secondary.

Even if the presence of the ossicle in question, as believed by Tornier, were limited to exceptional cases in one species of *Lepus*, the author's arguments would not be valid. It is one of the characteristics of these reduced " accessory " bones to ossify very late (Thilenius); and its exceptional appearance in *L. europæus* could, *a priori*, be interpreted quite as well in the sense of a disappearing element as in Tornier's sense. But the presence of this bone as a constant element in Lagomyidæ and several Leporidæ totally changes the aspect of the question. In the more primitive forms of Lagomorpha, the ossicle seems always to be present and proclaims itself a reduced element by its varying size and position. In those Leporidæ—of which *L. europæus* is the prototype—which are the most specialized for leaping, we must expect it to be of quite exceptional occurrence.

The ossicle has been observed in the " Hare " likewise by Pfitzner [||], who calls it the *præcuneiforme*. As to whether this and similar accessory bones are to be considered as " secondary " or " sesamoids," Pfitzner has shown us the way how to proceed [•], viz. that we cannot base our conclusions on the examination of a single specimen or a few species. The " præcuneiforme " has been studied by Pfitzner especially

---

G. Tornier. " Ueber den Saugetier-Praehallux. Ein dritter Beitrag zur Phylogenese des Saugetierfusses." Arch. f. Naturgesch. 1891. pp. 115–204.

† *Op. cit.* p. 181.   ‡ Meaning *Lepus europæus*, Pall.   § *Op. cit.* p. 196.

|| Morph. Arb. (Schwalbe) i. p. 533 (1892); iv. p. 354 (1895). Prof. Pfitzner has kindly informed me that the species is *L. europæus*, Pall.

• *Ll. cc.* ; and Morph. Arb. vi. p. 394 (1896).

in the Carnivora (where it had been seen by Meckel); and of the Polecat alone he examined seventeen specimens. His conclusions are summed up in the following words :—"Skelet und Musculatur variieren unabhängig von einander, da findet kein Ineinandergreifen beider Processe statt, höchstens, und stets nur in beschränktem Maasse, ein gewisses Nebeneinanderherlaufen. . . . Muskeln und Muskelansätze und Skelet variieren ohne innere Korrelation, und deshalb ist es für die Deutung eines bestimmten Skeletstückes ganz irrelevant, ob ein bestimmter Muskel sich daran ansetzt oder nicht. Das Praecuneiforme bleibt das Praecuneiforme, ob sich M. tib. anticus oder M. tib. posticus ganz, theilweise, oder gar nicht daran ansetzt, und für die Deutung eines Skeletstücks oder selbst seiner Komponenten, also für die etwaige Frage, ob andere Skeletstücke durch Assimilation mit ihm vereinigt sind, bleibt es ganz gleichgiltig und ohne jede Beweiskraft, welcher Muskel an ihm inseriert.—Aber auch mit den Bändern steht es nicht anders; auch sie variieren nach Vorkommen und Ausbildung ohne Rücksicht auf die Skeletstücke " *. And Thilenius : "Die Beziehungen, welche die accessorischen Elemente der chiropterygialen Wirbelthiere zu Muskeln, Sehnen oder Bändern besitzen, sind nicht primäre Erscheinungen, sondern secundär während der Ontogenese erworben " †.

When the "tibiale" is not a separate bone, as in many Rodents. it is considered to be part of the navicular, the "tuberositas navicularis medialis " (Baur, Leboucq. Emery ‡). It does not seem to me to preclude the assumption of a medial tibiale, which would be a part or the whole of Emery's "paracentrale" §, the first element of the second ray (II, 1). If then the tibiale marginale (or externum) is the first element of the first ray (I, 1), the suggestion lies not far off that, like the distal "praepollex," the distal "praehallux " (praecuneiforme) is the second element (I, 2) of the same ray, but that it has generally been thrust out of the series.

4. "*Accessory ossicles*" *articulating with Metatarsal V.*—On Pl. **38**, fig. 9, I have represented the enlarged figure of a right Metat. V. from La Grive, *a* from the dorsal, *b* from the volar side. This is still another instance of a fossil metapodial, presenting unusual articular facets. for which, for a long time. I was unable to account, for want of material for comparison.

The ossicle is much larger than the Metat. V of *Prolagus œningensis*, which otherwise resembles it closely, exhibiting the same particulars as do the fifth metatarsals of *Prolagus sardus* and *P. sardus* var. *corsicanus*. I must leave it undecided whether the figured metatarsal belongs to *Titanomys Fontannesi* or to *Lagopsis verus*, which, judged from other parts of their skeletons, were both of about the same size.

On the volar aspect (*b*) is seen a large facet. starting from the proximal end and running obliquely in the direction of the tuberositas lateralis. In Leporidæ I find in the

---

* Morph. Arb. vi. p. 394 (1896).

† *Ibid.* v. pp. 544, 545 (1895).

‡ C. Emery. "Beitr. z. Entwicklungsgesch. u. Morph. d. Hand- u. Fussskelets der Marsupialier." (Semon's 'Forschungsreisen in Australien,' ii. pp. 394, 395 (1897).

§ C. Emery. "Die fossilen Reste von *Archegosaurus* und *Eryops* und ihre Bedeutung für die Morphologie des Gliedmaassenskelets." Anat. Anz. xiv. pp. 206, 207, figs. 3–7 (1898).

corresponding part no facet, but instead, either a convex swelling of the region, or in some cases, on the contrary, a more or less rugose depression. In *Lagomys* (*L. rufescens* and *L. melanostomus*) there is the facet in the same place, and articulating with it a comparatively large orbicular or triangular ossicle. I think it probable that, in those Leporidæ (*Caprolagus hispidus*, *C. Netscheri*, *Sylvilagus brasiliensis*) where the corresponding region of the Metat. V is raised to a convex protuberance, the ossicle in question has become fused with the former bone.

A similar ossicle has been met with by Pfitzner in Carnivora, viz. in *Ursus arctos* and · in *Lutra* *. I find the same ossicle in Cercopithecidæ, in *Mus*, and, among Insectivora, in *Erinaceus*, *Gymnura*, *Myogale*, *Condylura* and Centetidæ (*Limnogale*, *Oryzoryctes*, *Microgale*). In the latter, and in *Myogale*, it is enlarged transversely and extends also on to the base of Metat. IV.

Pfitzner homologizes the ossicle in Carnivora with a similar one on the fifth metacarpal of Primates (os hamuli), and regards these and similar occurrences in the third tarsal ray (os unci, in *Felis*) as carpalia (or tarsalia) which have become " abortive," and have been secondarily displaced towards the volar side †. The question would then arise whether we have to consider the ossicle of the Metat. V as pertaining to the fifth or to the fourth ray; for from its position—and the same remark applies to the "os hamuli "— on the tibial side of the Metat. V, and sometimes articulating with Metat. IV also, it may belong to either. For the present the question cannot be settled; but since both tarsal and carpal elements in question are of apparently widespread occurrence, we may hope to get better acquainted with them before long. In the marsupial *Perameles obesula*, Metat. IV and Metat. V have each a separate plantar bone, articulating with their proximal capitulum ‡.

On the dorsal side of the tuberosity of Metat. V—on the left in fig. 9 *a*—is seen what appears to be a facet, partially extending to the volar side also. The same facet is present in both species of *Prolagus*. It at once recalls to mind what occurs on the Metac. V of *Prolagus* and *Lagomys*, and some Leporidæ, where carpale 5 (os vesalianum carpi) articulates with the tuberosity.

A distinct os vesalianum tarsi (Pfitzner) is a very rare occurrence in Man, in whom it has been seen by Vesalius, Gruber, and Spronck §. Pfitzner never saw it free; when distinct—one case figured by Vesalius, two described and figured by Gruber, one by Spronck—it is situated on the fibular side of the pes, " in the angle between the cuboid and Metat. V, articulating with both." An epiphysis which may occur on the tuberosity

---

* Tageblatt der 60. Vers. dentsch. Naturf. und Aerzte in Wiesbaden, p. 251 (1887).—Speaking of the Bear, the author states that the ossicle occurs on the plantar base of Metacarpal V ; from the context it would appear that this is a misprint for Metatarsal V · at any rate, in *Lutra* it is present on both Metacarpal and Metatarsal V, as stated by the same author.

† Morph. Arb. i. pp. 7, 8 (1891) ; 541, 542, 587 (1892) ; iv. p. 539-543 (1895).

‡ C. Emery, " Beitr. z. Entwicklungs-gesch. u. Morph. d. Hand- u. Fussskeletts der Marsupialia " (Semon's ‘Forschungsreisen in Australien,’ &c., ii. p. 381, Taf. xxxv. figs. 45, 46 (1897).

§ Morph. Arb. i. pp. 522, 595, 596, 756, 757 (1892); vi. pp. 472-475 (1896).

of the human fifth metatarsal is considered by Pfitzner * as one of the terminal stages of its assimilation by the latter bone.

I find an epiphysis on the tuberosity of Metat. V in the Rodent genera *Georychus* and *Clenomys*. The bone itself I have never seen independent, but, from what I have stated above as to the fossil metatarsal, there can hardly be a doubt that an ossicle articulates with the tuberosity. The cuboid of *Prolagus*, of which I have a number of specimens, shows a facet—absent in Leporidæ—on the proximal fibular side; and this, together with the facet on the tuberosity of Metat. V, suggests the presence in these Lagomyidæ of either one ossicle articulating distad with the Metat. V and proximad with the cuboid, or two ossicles, the proximal of the two articulating with the cuboid, the distal with Metat. V; both possibly articulating originally also with each other at their apposed surfaces. Considering the rather considerable distance which must have occurred between the two facets, the latter hypothesis—of two bones—seems the more probable.

The presupposed proximal ossicle would be the homologue of the "os peroneum" (Pfitzner) of Man † and other Primates, which is the so-called sesamoid in the terminal tendon of the peroneus longus muscle. It has in Man, according to Pfitzner, a frequence of about 8-9 °₀, and is situated on the postero-lateral end of the eminentia obliqua cuboidei. "Hier findet sich in den Fällen bester Ausbildung eine scharf abgesetzte Facette, der eine gleiche auf dem Peroneum entspricht " ‡. This os peroneum was seen by Daubenton in *Hylobates*: "Il y a de plus dans le gibbon un huitième os placé au côté externe du tarse, à l'endroit où le calcaneum touche au cuboïde " §. In the skeleton of a *Hylobates lar* in the Natural History Museum, there is to be seen an ossicle articulating with the cuboid; and it is of quite general occurrence among the Cercopithecidæ. Gillette mentions it in Monkeys generally as articulating with the cuboid ‖. Whether the ossicle mentioned by G. Fischer in the *Tarsius* is a vesalianum or a peroneum I cannot decide for the present. He says: "Auch findet man in den *Tarsern* noch ein überzähliges Beinchen, rund, linsenförmig, doch länglich, welches eigentlich auf dem letzten Mittelfussknochen aufsitzt, der sich immer mit seinem Kopfe weit nach hinten zieht "¶. I see the "peroneum" in a minute ossicle in *Limnogale* (an aquatic member of the Malagasy Centetidæ), adhering to the tendon of the musc. peroneus longus, laterally from the cuboid, and I believe the reason that it has not been more frequently seen in Mammals is that the muscle is generally cut away in preparing the skeleton.

* Morph. Arb. vi. pp. 262, 263, 474 (1896).

† *Ibid.* i. pp. 530, 531, 594-598, figs. 12, 13 (1892); vi. pp. 456-462 (1896).

‡ *Ibid.* vi. p. 456.

§ Buffon-Daubenton, Hist. nat. gén. et partic. xiv. p. 106 (1766).

‖ "Chez les singes, l'os sésamoïde du péronier latéral est très-volumineux, puisque, chez des individus de petite taille, nous l'avons trouvé au moins aussi gros que ceux du pouce de l'homme, constant et ayant la forme d'un trois-quarts d'ovoïde régulier; il possède une face véritablement articulaire, un peu convexe, et qui répond à une facette également encroutée de cartilage de la partie inférieure du cuboïde." (Journ. de l'Anat. et de la Physiologie, viii. p. 533, 1872.)

¶ Gotthelf Fischer, 'Anatomie der Maki,' p. 151 (1804). This ossicle is not mentioned in Burmeister's 'Beiträge z. näh. Kentniss der Gattung *Tarsius*' (1846).

The "peroneum" would then be homodynamous with the ulnare externum Thil. (ulnar part of Pfitzner's triquetrum bipartitum),$=$V 2; and the vesalianum tarsi with the vesalianum carpi,$=$V 3, or tarsale 5, the cuboid being tarsale 4.

The cuboid of Mammalia is generally considered to be a compound of tarsale 4+tarsale 5 ; but where an os vesalianum, or its traces on the tuberosity of Metat. V, are present such a supposition cannot, however, be admitted. Emery found in embryos and pouch specimens of the Marsupial genera *Didelphys*, *Epyprymnus*, and *Phascolarctus* separate tarsalia 4 and 5 *. For the former genus at least he has demonstrated that tarsale 4 and tarsale 5 become fused in later stages. This instance of a compound Mammalian cuboid (tarsale 4 and 5) is the only one in the literature which can be taken seriously; but it is quite possible that in other Mammalia too the vesalianum may be assimilated by the cuboid, instead of by Metat. V, as in Man and some Rodentia.

Concluding Remarks and Suggestions as to Classification.

The oldest known lagomorphine genera, *Titanomys*, and *Palæolagus*, have several important characters in common : still, the differentiation into Lagomyidæ (*Titanomys*) and Leporidæ (*Palæolagus*) had already taken place. In the number of upper molars and in the shape and composition of the bony palatal bridge, *Titanomys* shows itself the precursor of the recent *Lagomys*, *Palæolagus* of the recent *Lepus* ; and it is therefore advisable to retain the two groups as families, although they converge back in time. Moreover, in other characters—absence of supraorbital processes, pattern of the cheek-teeth—*Palæolagus* approaches nearer the Lagomyidæ than do the more recent Leporidæ. In the gradual transformation of their cheek-teeth, both groups, as has been amply demonstrated, run parallel from the Lower Miocene down to recent times. The Lagomyidæ, as at present known, start from a more primitive type than the Leporidæ, since in *Titanomys* the cheek-teeth have remnants of roots and the upper ones preserve their original pattern throughout life ; whereas in *Palæolagus*, so far as I know, the cheek-teeth are already rootless, and in old age they lose their original pattern, without, however, developing the new one. In the transformation of their tooth-pattern the Leporidæ eventually go a step beyond the point at which the Lagomyidæ stop, the cheek-teeth of *Lepus* being more completely metamorphosed than those of recent *Lagomys*. In this respect, as well as in the specialization of their limbs for swiftness, correlated with the greater perfection of the sense-organs—and, as a consequence, with corresponding modifications of the skull—the Leporidæ are to be considered the more specialized of the two ; but there are several members of the Leporidæ which, with regard to the two last-mentioned sets of characters, and the complete or almost complete absence of the tail, preserve considerable similarity to the Lagomyidæ. By the absence of the upper m. 3, and by some peculiarities of the cranium, pointed out by Winge (perforation of the fossa pterygoidea—fusion of the

---

* Atti Acc. Lincei, Rend. iv. 2, p. 274 (1895) ; id. in Semon's ' Forschungsreisen in Australien,' &c., ii. pp. 374, 378, 383 ; figs. 29, 30, 31, 59 (1897).

spongiose os tympanicum with the petrosum), the Lagomyidæ are more modernized than the Leporidæ.

If we take for a guide the gradual metamorphosis of their upper cheek-teeth, the order of succession in Lagomyidæ is : *Titanomys*, *Prolagus*, *Lagopsis*, *Lagomys*. *Lagomys* is clearly the offspring of *Lagopsis* ; but *Lagopsis* cannot be descended from *Prolagus*, the latter having lost the last lower molar (m. 3), which is present in *Lagopsis* (and *Lagomys*). *Lagopsis* must have taken its origin from a form with upper cheek-teeth like or nearly like those of *Prolagus*, but provided with a lower m. 3, a hypothetical "*Prolagopsis*" descended from *Titanomys* or some closely related form with persistent lower m. 3. In *Titanomys* (*T. eisenveiensis*) there is already the beginning of the tendency to the loss of this tooth. *Prolagus* equally descends from a *Titanomys*-like form, and has continued without much change from the Middle Miocene to the present era, since it still lingered in Corsica at the Neolithic period.

Titanomys

(*Prolagopsis*)   *Prolagus*

|

Lagopsis

|

Lagomys

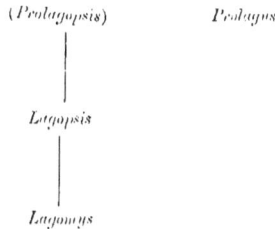

Leporidæ.—Apart from attempts to separate the Rabbit as a genus from the rest of the Leporidæ, which have not, however, met with common assent, the family has pretty generally been considered to be composed of one recent genus only, *Lepus*. In 1845, Blyth proposed a new genus *Caprolagus*, for Pearson's *Lepus hispidus* [*]. The appropriateness of this generic distinction has been contested by Hodgson and by Waterhouse. The former, omitting to take into consideration the remarkable configuration of the skull of the Hispid Hare, pointed out, that " In the Timid and Red-tailed Hares the long ears, the large eyes, the frame as well suited to extreme speed as the eyes and ears to effective vigilance, are certainly in remarkable contrast with the small eyes and ears, heavy frame, and short equal legs of the Forest Hare : but all these distinctions, as well as those of domicile, become less and less tangible in the Variable Hare, the Rabbit, the Tolai and the Tapiti, in which moreover we have variously reproduced, even to the subordinate peculiarities of the Indian Forest Hare, such as its white flesh, its short tail, its subterranean retreat and creeping adhesion thereto, so unlike the dashing career of the

---

[*] E. Blyth. "Description of *Caprolagus*, a new Genus of Leporine Mammalia ; with two plates." Journ. As. Soc. Bengal, xiv. i. pp. 247-249 (1845).

red-tailed and English species " *.   Waterhouse's objections are to the following effect :—
"Strongly marked . . . as these distinctions are, if the Assam Hare be compared with
the Common Hare, they are less so when that animal is compared with the Indian Hare
(*Lepus ruficaudatus*), and *much* less so when it is compared with the *Lepus brachyurus*
of Japan.   This last-mentioned animal has the short ears and tail of the *Lepus hispidus*,
and the same large molar and incisor teeth, combined with a powerfully-formed skull,
but in this skull the notch which sets free from the fore part of the supraorbital process
is not absent, as in *Lepus hispidus*: it agrees in having the patch unusually long, but
differs from the skull of *L. hispidus* (as it would appear from Mr. Blyth's figures) in
having the zygomatic arches straight and parallel as in other Hares ; the Assam species
having the zygoma somewhat arched outward.   The peculiarities which I have pointed
out as distinguishing the lower jaw of the *Lepus ruficaudatus* from that of the *L. timidus*
are also found in the lower jaw of *L. hispidus*, but here the angular portion has a still
greater transverse diameter " †.   The result of these criticisms was the withdrawal of
the genus *Caprolagus* by its author ‡.

For my part, I am unable to accept these opinions.   Some of the remarks of the
former writers are undoubtedly just, and two of the examples of other Leporine species,
adduced by Hodgson, as resembling the Hispid Hare, are more to the point than
Waterhouse's comparisons.   But the conclusions I infer from them are very different
from those of these authors.   The external characters and the conformation of the skull
and limbs, in which the Hispid Hare is distinguished from *L. europæus*—taking this latter
as the type of the genus *Lepus* s. str.—are very remarkable.   The circumstance, which
I shall more fully point out hereafter, that there are other Leporines approaching the
Hispid, simply shows that the latter—apart from its specialization as the only true
fossorial member of the family—does not stand alone, and that several other species
equally deserve to be separated from the genus *Lepus*.

The first attempt at a tabular arrangement of the species of *Lepus*, according to their
affinities, was made by Baird §, who availed himself of the characters of the skull ; limiting
himself—with the exception of " *Lepus cuniculus* "—to North American species.   The six
sections into which the genus is divided show that this excellent observer had on the
whole a right conception of the affinities of this group.   Not all his sections, however,
are of equal value ; section B, comprising *L. californicus* and *L. callotis*, is in reality
more closely related to A (*L. timidus*, *L. glacialis*, *L. americanus*, *L. campestris*, &c.)
than to the other sections ; and the same may be said with regard to E (*L. Trowbridgei*
and *L. Audubonii*), which, as a matter of fact, is in closer relation with D (*L. sylvaticus*
and allies) than with the rest.

With such a good example to follow, a successor, taking up the whole of the known
Leporidæ, might have been enabled to make a further step forward.   This is what J. E.

* B. H. Hodgson, " On the Hispid Hare of the Saul Forest " (Journ. As. Soc. Bengal, xvi. i. p. 574 (1847).
† G. R. Waterhouse, 'A Natural History of the Mammalia,' ii. p. 80 (1848).
‡ E. Blyth, Catal. Mamm. in Mus. Asiat. Soc. Calcutta, p. 133 (1863).
§ Spencer F. Baird, 'Mammals of North America,' pp. 574, 575 (1859).

Gray attempted to do *. From the title of the article, "Notes on the Skulls of Hares (Leporidæ) and Picas (Lagomyidæ) in the British Museum," the actual contents could not be guessed, for the work is an attempt at a complete classification of the Lagomorpha, in which several characters besides cranial are made use of. The characters assigned to the family Leporidæ are in part either erroneous (characteristics of the molars), or they do not hold good for all the minor divisions, and are consequently partly in contradiction with the subsequent characteristics of the sections. This family is divided first of all into two sections, one reserved for Blyth's *Caprolagus*, the other for the rest of the Leporidæ. This latter is again subdivided into two groups:—A. Hares, B. Rabbits, the latter containing the Rabbit alone, raised to generic rank (*Cuniculus*). In group A are given generic names to some of Baird's divisions. The latter's D (ex *L. sylvaticus*) becomes *Sylvilagus*, his F (*L. aquaticus, L. palustris*) *Hydrolagus*; while a genus *Tapeti* is created for the Brazilian Hare, and *Eulagos* for "*L. mediterraneus*" and "*L. Judææ*." In the subdivisions of this A group (Hares), great stress is laid upon a comparatively unimportant cranial character, which had cautiously been made use of by Baird. Thus we get two subdivisions: I. Postorbital process more or less united with the skull (*Hydrolagus, Sylvilagus, Eulagos*). II. Postorbital process separate from the skull (*Lepus, Tapeti*).

The species of the genus "*Lepus*" are classed according to geographical distribution, and thus there are unavoidably thrown together very heterogeneous forms in the African, Asiatic, and American members. Among the latter are *L. Audubonii* and *L. Trowbridgei*, which are thus widely separated from *Sylvilagus*, containing their closest allies.

The fore-mentioned paper was wisely ignored by J. A. Allen, in his Analysis of the species and varieties of North American Leporidæ †. Allen on the whole follows Baird, with some improvements in detail, but with one step backward, by widely separating the *Callotis* group from *L. timidus* and its allies.

Some of Gray's generic names have since been used as subgenera, e. g. by Mearns, with whose "Analysis of three Subgenera of *Lepus*" ‡, containing some valuable information, I propose to deal elsewhere.

A new genus of Leporidæ, *Romerolagus*, from Mount Popocatepetl (3350 metres), was described some years ago by Hart Merriam §. The author's views as to its systematic position are summed up in the following words :—" The skull, singularly enough, does not show the departure from *Lepus* that one would expect from a study of the other bones. It agrees in the main with skulls of the American Cottontails (subgenus *Sylvilagus*), but differs in the postorbital processes, which are small, divergent posteriorly, and altogether wanting anteriorly, and in the jugal, which is greatly elongated posteriorly. The interparietal is distinct, and in old age becomes ankylosed with the supraoccipital. The thoroughly leporine character of the skull shows that the animal can hardly be regarded as ancestral to *Lepus*, as might have been

* Ann. & Mag. Nat. Hist. xx. 3, p. 219 (1867).
† 'Monographs of North American Rodentia.—II. Leporidæ,' by J. A. Allen, p. 283 (1876).
‡ Proc. U. S. Nat. Mus. xviii. p. 551 (1896).
§ Proc. Biol. Soc. of Washington, x. p. 169-174 (1896).

inferred from its short ears, short hind legs, and various skeletal characters, but that it is a specialized offshoot from the genus *Lepus* itself " *.

My own views as to the significance of the characters of *Romerolagus* are about the same as those with regard to *Caprolagus*. They are certainly of generic value, by comparison with those generally assigned to the genus *Lepus*. But it does not follow that *Romerolagus* can stand as a separate genus, or, to put it in a more general way, that it occupies an isolated position compared with other Leporidæ. I feel sure that if the same care had been bestowed on the examination of the skeletons of some other Leporidæ near at hand, *e. g.* the aquatic Hares †, Hart Merriam would have arrived at the same conclusion as I have. It will probably be possible to show hereafter that *Romerolagus is* specialized in some respects, as might be anticipated from its habitat. The remarkable shortness of the ears is presumably the combined result of inheritance and specialization. The absence of the tail is certainly an acquired character, as it is in *Lagomys*. The complete clavicle can scarcely be regarded in the same light; but, although I know of no other member of the Leporidæ having a " complete " clavicle, *Romerolagus* does not, in this respect either, occupy such an isolated position as the author seems to think. That the skull is "thoroughly leporine" I cannot admit; there are several cranial characters, as will be shown, which are unusual in most Leporidæ, but which *Romerolagus* shares with *Palæolagus*, with some recent Leporidæ, and with the Lagomyidæ, and which may be regarded as ancestral.

The whole of recent Leporidæ may be divided into two groups, probably of higher than generic dignity, which might conveniently be called : A. *Caprolagus* group, and B. *Lepus* group.

A. *Caprolagus* Group :—

    1. *Caprolagus*: *C. sivalensis*, Maj.; *C. raldarnensis* (Weith.); *C. hispidus* (Pears.) (type.)

    2. *Nesolagus* (nom. nov.) : *N. Netscheri* (Schleg. & Jent.).

    3. *Oryctolagus*: *O. cuniculus* (Linn.); *O. crassicaudatus* (Geoffr.).

    4. *Sylvilagus*, comprising in this term :—

        a. *Limnolagus* (*S. palustris, aquaticus,* &c.).

        b. *Romerolagus* (*S. Nelsoni*, Merr.).

        c. *Tapeti* (*S. brasiliensis,* &c.).

        d. *Sylvilagus* (*S. sylvaticus,* &c.)

The question whether 1–4 are to be considered as genera or subgenera is for the present quite immaterial. *Sylvilagus* s. str. is the least primitive of this group, and *Oryctolagus* stands somewhat apart.

B. *Lepus* Group.—This group contains the one genus *Lepus* s. str., including all the species not contained in group A.

* *Op. cit.* p. 172.

† This remark refers also to the limbs, although I do not know them from either.

The *Caprolagus* group (A) differs from the *Lepus* group by the following characters, part of which, as said above, it shares with *Palæolagus* and with the Lagomyidæ : —

Lesser specialization for speed, and in correlation with this, lesser development of organs of sense (sight, smell, hearing). Fore and hind feet comparatively short and subequal. Ears short. Eyes smaller. Tail very short or missing.

Cranium, depressed above, anteriorly and posteriorly very little bent downward. Upper contour of frontals and posterior part of nasals almost horizontal (exc. *Oryctolagus*). Inferior border of orbit—formed by malar bone—shorter than in the *Lepus* group : sinus on the lateral face of malar not advancing so far forward (exc. in *Oryctolagus*). Upper border of zygoma bent inward, inferior border arched outward (exc. in *Oryctolagus*). Posterior appendix of zygoma elongate and, in correlation, mandibular condyloid process elongate also (exc. in *Oryctolagus crassicaudatus*).

Infraorbital foramen larger than in *Lepus* and its immediate neighbourhood almost devoid of reticulation. The heavier skull in the A group is in evident correlation with the different mode of locomotion. The following cranial characters of A are apparently in correlation with the less developed organ of smell :—Horizontal portion of os palatinum comparatively well developed ; interpterygoid fossa and choanæ comparatively small. Foramina incisiva comparatively narrow and short. Anterior part of nasals less inflated than in *Lepus*. In correspondence with the smaller eyes, the orbits are comparatively small, and the orbital processes more or less reduced.

In conclusion, I wish to express my very special obligations to Prof. Howes for loan of material, valuable suggestions, and the pains he has taken in connection with this memoir.

EXPLANATION OF THE PLATES.

PLATE 36.

Fig. 1. *Caprolagus (Oryctolagus) cuniculus* (Linn.), juv. Right maxillary ; d. 3–m. 2.

Fig. 2. *Plesiadapis Daubrei*, Lem. Right upper molar. Enlarged copy from Bull. Soc. Géol. France, 3. xix. (1891) pl. x. fig. 62 *u*.

Fig. 3. *Pelycodus helveticus*, Rüt. Right upper molar. Enlarged copy from Abh. Schweiz. Pal. Ges. xv. pl. fig. 12 *a* (1888).

Fig. 4. *Prolagus sardus* (Wagn.). Left maxilla with deciduous teeth (d. 3–d. 1) and first molar. Monte San Giovanni (Sardinia). Br. Mus. G. D. No. M3464.

Fig. 5. *Caprolagus (Oryctolagus) cuniculus* (Linn.) ; slightly older than fig. 1. Right maxilla ; d. 3–m. 2 ; alveolus of m. 3.

Fig. 6. *Titanomys Fontannesi* (Dep.). Second (last) right upper molar (m. 2), almost disused. Middle Miocene. La Grive-Saint-Alban (Isère), as all the other specimens of this species *.

Fig. 7. *Titanomys Fontannesi* (Dep.). First right upper molar (m. 1).

Fig. 8. *Titanomys Fontannesi* (Dep.). Posterior right upper premolar (p. 1). Br. Mus. G. D. No. 5268

Fig. 9 *Titanomys Fontannesi* (Dep.) ? Second right upper premolar (p. 2) ? *

Fig. 10. *Prolagus œningensis* (Kön.). The three left upper premolars (p. 3–p. 1) of young specimen. Middle Miocene. La Grive-Saint-Alban, as all the other specimens of this species. Br. Mus. G. D. No. 5234.

* The figures marked thus are from specimens in the possession of the author.

71*

Fig. 11. *Prolagus sardus* (Wagn.). Posterior right upper premolar (p. 1), from a young specimen. Monte San Giovanni (Sardinia). Br. Mus. G. D. No. M3461.

Fig. 12. *Titanomys Fontannesi* (Dep.). First left upper molar (m. 1), slightly worn.

Fig. 13. *Titanomys Fontannesi* (Dep.). First left upper molar (m. 1), slightly worn.

Fig. 14. *Titanomys Fontannesi* (Dep.). Right upper, probably deciduous, molar ; much worn *.

Fig. 15. *Titanomys Fontannesi* (Dep.). Posterior right upper premolar (p. 1) *.

Fig. 16. *Prolagus sardus* (Wagn.). Fragment of right maxillary ramus, with posterior premolar (p. 1), and the two true molars (m. 1, m. 2). Monte San Giovanni. Br. Mus. G. D. No. M3459.

Fig. 17. *Caprolagus (Oryctolagus) cuniculus* (Linn.). Young individual, slightly older than fig. 5. The two posterior premolars (p. 2, p. 1) and the two anterior molars (m. 1, m. 2) of the right side.

Fig. 18. *Titanomys risenoviensis*, H. v. Mey. Upper molar, much worn. Bravard Collection. Lower Miocene, Allier. Br. Mus. G. D. No. 31094–104.

Fig. 19. *Titanomys risenoviensis*, H. v. Mey. The two posterior premolars (p. 2, p. 1), from a fragment of the right maxillary. Lower Miocene of Weisenau (Germany). Br. Mus. G. D. No. 21495.

Fig. 20. *Caprolagus (Sylvilagus) brasiliensis* (Linn.). Right upper posterior deciduous molar (d. 1), from a skull in the Br. Mus. Z. D. No. 58.4.15.1.

Fig. 21. *Prolagus œningensis* (Kön.). Complete series of the right upper cheek teeth (p. 3–m 2) *.

Fig. 22. *Lepus timidus*, Linn. (*L. variabilis*, Pall.). Right upper cheek-teeth of young individual ; the two posterior deciduous molars have been removed, in order to show the overlying premolars (p. 2, p. 1). Ireland. W. E. de Winton, Esq.

Fig. 23. *Titanomys Fontannesi* (Dep.). Left upper jaw, showing the empty alveoli of the five cheek-teeth. 4 × 1 *.

Fig. 24. *Prolagus sardus* (Wagn.). Complete series of the right upper cheek-teeth, or (p. 3–m. 2). Ossiferous breccia of Monte San Giovanni (Sardinia) *.

Fig. 25. *Lepus europeus*, Pall. Unworn right upper median premolar (p. 2) of young individual. From a skull in the Br. Mus. Z. D. No. 523 l.

Fig. 26. *Lepus timidus*, Linn. Posterior right upper deciduous molar (d. 1), removed from the jaw fig. 22.

Fig. 27. *Caprolagus hispidus* (Pears.). Median right upper premolar (p. 2), of young individual in the Br. Mus. Z. D.

Fig. 28. *Lepus* sp. Right upper deciduous molars (d. 3–d. 1). China. Br. Mus. Z. D.

Fig. 29. *Prolagus œningensis* (Kön.). Right upper deciduous molar (either d. 1 or d. 2) *.

Fig. 30. *Lagopsis verus* (Hens.). Right upper deciduous molar (either d. 1 or d. 2). Middle Miocene of La Grive-Saint-Alban *.

Fig. 31. *Lagopsis verus* (Hens.). Median right upper premolar (p. 2). La Grive-Saint-Alban. Br. Mus. G. D.

Fig. 32. *Lagopsis verus* (Hens.). Left upper molar. La Grive-Saint-Alban. Br. Mus. G. D.

Fig. 33. *Caprolagus hispidus* (Pears.). Complete series of the right upper cheek-teeth. Adult. From a skull in the Br. Mus. Z. D.

Fig. 34. *Lepus nigricollis*, F. Cuv. Posterior right upper premolar (p. 3). Br. Mus. Z. D. No. 81.4.29.7.

Fig. 35. *Titanomys Fontannesi* (Dep.). Left upper deciduous molar (either d. 1 or d. 2) *.

Fig. 36. *Palæolagus Haydeni*, Leid. Fragment of right maxillary ramus, showing the empty alveolus of the median premolar (p. 2), and the three following cheek-teeth (p. 1, m. 1, m. 2). Br. Mus. G. D. No. M5727.

PLATE 37.

Fig. 1. *Titanomys Fontannesi* (Dep.). Isolated lower anterior premolar (p. 2), unworn. Middle Miocene of La Grive-Saint-Alban, like all the other specimens of this species *.

Fig. 2. *Titanomys Fontannesi* (Dep.). Another isolated specimen of the same tooth, slightly worn *.

Fig. 3. *Titanomys Fontannesi* (Dep.). A third isolated specimen of the same, slightly more worn by attrition than the former *.

Fig. 4. *Titanomys Fontannesi* (Dep.). A fourth isolated specimen of the same, much worn *.

Fig. 5. *Prolagus sardus* (Wagn.), var. *corsicanus*. The two inferior deciduous molars (d. 2, d. 1) from a left mandibular ramus. The first true molar (m. 1) of the specimen, not figured, shows a vestige of the terminal cusp ("*t* "=hypoconulid).

Fig. 6. *Prolagus œningensis* (Kön.). Left mandibular ramus of young individual, showing the two deciduous (d. 2, d. 1) and the two true molars (m. 1, m. 2). La Grive-Saint-Alban. Br. Mus. G. D. No. M5236.

Fig. 7. *Titanomys Fontannesi* (Dep.). Complete series of the lower cheek-teeth (p. 2–m. 3) in a left mandibular ramus. Adult. Br. Mus. G. D. No. 5267 *a*.

Fig. 8. *Caprolagus* (*Oryctolagus*) *cuniculus* (Linn.). The two lower premolars (p. 2, p. 1), in a right mandibular ramus of an immature specimen. Herefordshire. W. E. de Winton, Esq.

Fig. 9. *Prolagus œningensis* (Kön). Complete series of inferior cheek-teeth (p. 2–m. 2), left side. Adult *.

Fig. 10. *Titanomys Fontannesi* (Dep.). Posterior premolar (p. 1) and anterior true molars (m. 1, m. 2) in a left mandibular ramus of an immature specimen. p. 1, being still in the socket, has not yet come into attrition. Br. Mus. G. D. No. 5267 *b*.

Fig. 11 *a*. *Titanomys visenoviensis*, H. v. Mey. Isolated upper posterior premolar (p. 1), or anterior molar (m. 1?, right side. Lower Miocene of Weisenau. Br. Mus. G. D. No. 7217 *c*.

Fig. 11 *b*. *Titanomys visenoviensis*, H. v. Mey. Probably posterior premolar (p. 1) or anterior molar (m. 1). Left side. Lower Miocene of Weisenau. Br. Mus. G. D. No. 7217 *d*.

Fig. 12. *Prolagus œningensis* (Kön.). Anterior premolar (p. 2) from a left mandibular ramus.

Fig. 13. *Caprolagus Lacosti* (Pomel). Anterior premolar (p. 2) from a left mandibular ramus. Upper Pliocene of Perrier (France). Br. Mus. G. D. No. 27618.

Fig. 14. *Lagopsis verus* (Hens.). The two posterior true molars (m. 2, m. 3) from a right mandibular ramus. La Grive-Saint-Alban. Br. Mus. G. D. No. 5263.

Fig. 15. *Titanomys Fontannesi* (Dep.). Upper view of left mandibular ramus, exhibiting the empty alveoli of the five cheek-teeth *.

Fig. 16. *Titanomys Fontannesi* (Dep.). Posterior premolar (p. 1) and the two anterior true molars (m. 1, m. 2) in a right mandibular ramus. Empty alveoli of p. 2 and m. 3. Br. Mus. G. D. No. M5267 *c*.

Fig. 17. *Caprolagus* (*Nesolagus*) *Netscheri* (Schleg. & Jent.). Posterior upper premolar (p. 1), right side, from the skull in the Br. Mus. Z. D. No. 92.5.24.1. Sumatra.

Fig. 18. *Caprolagus sivalensis*, Maj. The two inferior premolars (p. 2, p. 1), from a fragment of the left mandibular ramus. Pliocene, Siwalik Hills, India. Cautley Coll. Br. Mus. G. D. No. 16529. (By an inadvertence of the artist, the anterior side of the teeth is turned to the right—their outer side being directed upwards in the plate—instead of to the left, as in all the other figures of mandibles and teeth of the left side.)

Fig. 19. *Caprolagus* (*Romerolagus*) *Nelsoni* (Merr.). Anterior premolar (p. 2), from a right mandibular ramus. Mount Popocatepetl (Mexico). Br. Mus. Z. D.

Fig. 20 *a, b*. *Caprolagus* (*Sylvilagus*) *palustris* (Bachm.). Lower anterior premolars (p. 2), right (*a*) and left (*b*), from a specimen in the Br. Mus. Z. D.

Fig. 21. *Titanomys Fontannesi* (Dep.). The posterior premolar (p. 1) and the two anterior true molars (m. 1, m. 2) in a fragment of the right mandibular ramus.

Fig. 22. *Caprolagus hispidus* (Pears.). The two premolars (p. 2, p. 1) from the right mandibular ramus of an immature specimen in the Br. Mus. Z. D.

Fig. 23. *Caprolagus hispidus* (Pears.). Complete series of lower cheek-teeth (p. 2–m. 3) from a right mandibular ramus of an adult specimen in the Br. Mus. Z. D.

Fig. 24. *Titanomys risenoriensis*, H. v. Mey.   The two anterior true molars (m. 1, m. 2) from a frag-
    ment of the left mandibular ramus.   Bravard Coll.   Lower Miocene of Allier (France).   Br.
    Mus. G. D. No. 31095.

Fig. 25. *Titanomys risenoviensis*, H. v. Mey.   The two premolars (p. 2, p. 1), from a fragment of the
    right mandibular ramus.   Lower Miocene of Allier.   Br. Mus. G. D. No. 31096.

Fig. 26. *Lagopsis verus* (Hens.).   The four anterior cheek-teeth (p. 2, p. 1, m. 1, m. 2) and the empty
    alveolus of the last (m. 3), in a left mandibular ramus.   La Grive-Saint-Alban *.

PLATE **38.**

Fig. 1. *Caprolagus* (*Sylvilagus*), sp., jun., from Bogotá.   Right manus, anterior or upper surface view.
    Nat. size.   *r*—vesalianum (carpale V) ; *h*—hamatum (carpale IV).   Nat. size.

Fig. 2. The same.   External (ulnar) view.   Nat. size.

Fig. 3. *Caprolagus* (*Oryctolagus*) *crassicaudatus* (Is. Geoffr.).   Br. Mus. Z. D.   No. 96.6.6.1.   Left
    manus, ulnar view.   Nat. size.

Fig. 4. *Lagomys rufescens*, Gray.   Br. Mus. Z. D.   Right manus, anterior view.   2 × 1.

Fig. 5. *Caprolagus* (*Oryctolagus*) *cuniculus* (Linn.), juv.   Right tarsus, ulnar view.   Nat. size.

Fig. 6. *Caprolagus* (*Sylvilagus*) *brasiliensis* (Linn.), juv.   Roy. Coll. Sc., London.   Right tarsus, ulnar
    view.   Nat. size.

Fig. 7. *Caprolagus* (*Sylvilagus*), sp. juv.   Bogotá.   Right tarsus, ulnar view.   Nat. size.

Fig. 8. *Caprolagus* (*Sylvilagus*) *brasiliensis* (Linn.), juv.   Roy. Coll. Sc., London.   Right antebra-
    chium and manus.   *a*, anterior, *b*, posterior or volar view.   Nat. size.

Fig. 9. *Lagopsis verus* (Hens.), or *Titanomys Fontannesi* (Dep.).   Middle Miocene, La Grive-Saint-Alban.
    Br. Mus. G. D. No. M5274.   Right metatarsus V.   *a*, anterior ; *b*, posterior view.   2 × 1.

Fig. 10. *Lagopsis verus* (Hens.), or *Titanomys Fontannesi* (Dep.).   Middle Miocene, La Grive-Saint-Alban.
    Br. Mus. G. D. M 5273.   Left ulna.   *a*, anterior ; *b*, posterior view.   Nat. size.

Fig. 11. *Prolagus sardus* (Wagn.).   Left ulna.   *a*, anterior ; *b*, posterior view.   Nat. size.   Pleistocene
    breccia, Monte San Giovanni (Sardinia).   Br. Mus. G. D. M 3471.

Fig. 12. *Prolagus sardus* (Wagn.).   Right radius.   *a*, anterior ; *b*, external (ulnar) ; *c*, internal (radial) ;
    d, posterior view.   Nat. size.   Pleistocene breccia, Monte San Giovanni (Sardinia).   Br. Mus.
    G. D. M3471.

Fig. 13. *Prolagus sardus* (Wagn.).   Left metatarsus II, from behind.   3 × 1.   Pleistocene breccia,
    Monte San Giovanni (Sardinia).   Br. Mus. G. D.

Fig. 14. *Lagomys rufescens*, Gray.   Left metatarsus II, external (fibular) view.   3 × 1.   Br. Mus. Z. D.
    No. 74.11.21.17.

Fig. 15. The same.   Posterior view.   3 × 1.

Fig. 16. The same.   Anterior view.   3 × 1.

Fig. 17. *Prolagus œningensis* (Kön.).   Left metatarsus II, anterior view.   3 × 1.   Middle Miocene,
    La Grive-Saint-Alban.   Br. Mus. G. D. No. M5248.

Fig. 18. The same.   External (fibular) view.

Fig. 19. *Prolagus sardus* (Wagn.).   Right metacarpal V, from the outer (ulnar) side, to show the facet for
    the os vesalianum (carpale V).   5 × 1.   Pleistocene breccia, Monte San Giovanni (Sardinia).
    Br. Mus. No. G. D. No. M3471.

Fig. 20. *Lagomys rufescens*, Gray jun.   Right antebrachium.   *a*, front view ; *b*, external (ulnar) ; *c*, internal
    (radial) view.   2 × 1.   Br. Mus. Z. D.

Fig. 21. *Lagomys rufescens*, ad.   Right radius.   *a*, internal (radial) ; *b*, front view.   2 × 1.   Khorassan.
    Col. Yate.

Fig. 22. *Caprolagus* (*Oryctolagus*) *crassicaudatus* (Is. Geoffr.).   Left tarsus and metatarsus ; internal
    (tibial) view.   Nat. size.   From skel.   Br. Mus. Z. D. No. 96.6.6.1.

Fig. 23. *Caprolagus* (*Nesolagus*) *Netscheri* (Schleg. & Jent.). Right tarsus and metatarsus; internal (tibial) view. Nat. size. From skel. Br. Mus. Z. D.†

Fig. 24. *Caprolagus* (*Caprolagus*) *hispidus* (Pears.). Right tarsus and metatarsus, internal (tibial) view. Nat. size. Br. Mus. Z. D.

Fig. 25. *Lepus timidus*, Linn. (*L. variabilis*, Pall.). Right tarsus and metatarsus, internal (tibial) view. Nat. size. Ireland. Br. Mus. Z. D. No. 76.4.10.2.

Fig. 26. *Lagomys rufescens*, Gray. Right tarsus and metatarsus, internal (tibial view). 2 × 1.

Fig. 27. *Prolagus sardus* (Wagn.), var. *corsicanus*. Left metatarsus II. *a*, anterior; *b*, internal; *c*, posterior; *d*, external view. Nat. size. Pleistocene breccia, Toga nr. Bastia (Corsica). Br. Mus. G. No. D. M 3486.

Fig. 28. *Caprolagus* (*Nesolagus*) *Netscheri* (Schleg. & Jent.). Left antebrachium. *a*, front view; *b*, external (ulnar) ; *c*, internal (radial) ; *d*, posterior view. Nat. size. Sumatra. Br. Mus. Z. S. Sumatra. Br. Mus. Z. D.

Fig. 29. *Lagomys rufescens*, Gray. Right ulna. *a*, external (ulnar) view (almost posterior in adult Leporidæ) ; *b*, posterior view (almost internal in adult Leporidæ); *c*, front view (almost external in adult Leporidæ). 2 × 1. Khorassan. Col. Yate.

Fig. 30. *Caprolagus* (*Oryctolagus*) *cuniculus* (Linn.). Left antebrachium. *a*, front; *b*, external (ulnar) ; *c*, internal (radial) ; *d*, posterior view. Nat. size. Herefordshire.

PLATE 39

Fig. 1. *Titanomys Fontannesi* (Dep.). m. 1, sup. dext. Anterior view. 3 × 1. Middle Miocene. La Grive-Saint-Alban *.

Fig. 2. *Titanomys Fontannesi* (Dep.). m. 2, sup. dext. *a*, anterior; *b*, external view. 3 × 1. La Grive-Saint-Alban *.

Fig. 3. *Sciuropterus fuscocapillus*, Blyth. m. 2, sup. dext. Anterior view. 4 × 1. Br. Mus. Z. D. No. 52.5.9.19.

Fig. 4. *Titanomys Fontannesi* (Dep.). Upper deciduous molar, much worn. Anterior view. 3 × 1. Tooth figured Pl. 36. fig. 14.

Fig. 5. *Titanomys visenoriensis*, H. v. Mey. p. 2, sup. sin. *a*, posterior; *b*, lower view. 4 × 1. Lower Miocene, Weisenau. Br. Mus. G. D. No. M7217.

Fig. 6. *Titanomys Fontannesi* (Dep.). Right lower molar. *a*, anterior; *b*, inner; *c*, outer view. 2½ × 1. La Grive-Saint-Alban *.

Fig. 7. *Caprolagus* (*Oryctolagus*) *cuniculus* (Linn.), juv. m. 1, sup. sin., posterior view. 3 × 1.

Fig. 8. *Caprolagus* (*Oryctolagus*) *cuniculus* (Linn.), juv. m. 1, sup. dext. Anterior view. 5 × 1. Specimen figured Pl. 36. fig. 1.

Fig. 9. *Caprolagus* (*Oryctolagus*) *cuniculus* (Linn.), juv. dec. 1, sup. dext. *a*, anterior; *b*, outer view. 5 × 1. Specimen figured Pl. 36. fig. 1.

Fig. 10. *Pteromys melanotis*, Gray. m. 2, sup. dext. Anterior view. 3 × 1. Br. Mus. Z. D. No. 48.8.15.2.

Fig. 11. *Titanomys visenoriensis*, H. v. Mey. Left lower molar. *a*, outer ; *b*, inner; *c*, anterior view. 2¼ × 1. Br. Mus. G. D. No. 21495.

Fig. 12. *Titanomys Fontannesi* (Dep.) ? p. 2 sup. dext. (?) Anterior view. 4 × 1. La Grive-Saint-Alban. Specimen figured Pl. 36. fig. 9. After renewed examination, the generic affinities of this tooth seem very doubtful.

† *pc.* of this fig. to be read together (= *pc.* præcuneiform).

520 FOSSIL AND RECENT LAGOMORPHA.

Fig. 13. *Titanomys Fontannesi* (Dep.). m. 1 sup. sin. Anterior view. 3 × 1. La Grive-Saint-Alban. Specimen figured Pl. **36**. fig. 12.

Fig. 14. *Titanomys risenoviensis*, H. v. Mey. m. sup. dext. Anterior view. 3 × 1. Specimen figured Pl. **36**. fig. 18.

Fig. 15. *Scinropterus xanthipes* (Milne-Edw.). m. 2, sup. dext. Anterior view. 3 × 1. Br. Mus. Z. D. No. 95.7.5.1.

Fig. 16. *Titanomys risenoriensis*, H. v. Mey. m. 2 sup. dext. Lower view. 10 × 1. Weisenau. Br. Mus. G. D. No. M7217.

Fig. 17. *Caprolagus (Sylvilagus) brasiliensis* (Linn.), juu. Sternum. Front view. Nat. size. Royal College of Science, London.

Fig. 18. *Caprolagus (Nesolagus) Netscheri* (Schleg. & Jent.). Sternum. Front view. Nat. size. Br. Mus. Z. D.

Fig. 19. *Titanomys Fontannesi* (Dep.). p. 1 sup. dext. Anterior view. 3 × 1. Specimen figured Pl. **36**. fig. 8.

Fig. 20. *Pteromys nitidus*, Desm. Germ of m. 2, sup. dext. Anterior view. 3 × 1.

Fig. 21. *Titanomys Fontannesi* (Dep.). dec. sup. sin. Anterior view. 3 × 1. Specimen figured Pl. **36**. fig. 35.

Fig. 22. *Lagopsis verus* (Hens.). dec. sup. dext. Anterior view. 4 × 1. Specimen figured Pl. **36**. fig. 30.

Fig. 23. *Prolagus œningensis* (Kön.). dec. sup. dext. Anterior view. 7 × 1. La Grive-Saint-Alban. Specimen figured Pl. **36**. fig. 29.

Fig. 24. *Prolagus sardus* (Wagn.), var. *corsicanus*. Metatarsus II. Pleistocene breccia of Toga, near Bastia (Corsica). 3 × 1.

Fig. 25. *Titanomys Fontannesi* (Dep.). Left mandibular ramus. *a*, inner; *b*, outer view. Nat. size. La Grive-Saint-Alban *.

Fig. 26. *Prolagus œningensis* (Kön.). dec. 2 inf. 4 × 1. La Grive-Saint-Alban *.

Fig. 27. *Prolagus sardus* (Wagn.), var. *corsicanus*. Right mandibular ramus. *a*, inner; *b*, outer view. Nat. size. Pleistocene breccia, Toga (Corsica) *.

Fig. 28. *Caprolagus (Nesolagus) Netscheri* (Schleg. & Jent.). Right mandibular ramus, outer view. Nat. size. Br. Mus. Z. D. No. 92.5.24.1.

Fig. 29. *Titanomys Fontannesi* (Dep.). Posterior fragment of right mandibular ramus. *a*, outer; *b*, inner view. Nat. size. La Grive-Saint-Alban *.

Fig. 30. *Lagopsis verus* (Hens.). Right mandibular ramus. *a*, outer; *b*, inner view. Nat. size. La Grive-Saint-Alban *.

Fig. 31. *Titanomys Fontannesi* (Dep.). Left mandibular ramus. *a*, outer; *b*, inner view, Nat. size. La Grive-Saint-Alban *.

Fig. 32. *Caprolagus (Caprolagus) hispidus* (Pears.). Palatal view of skull. Nat. size. Br. Mus. Z. D. No. 48.9.12.11. *m*=maxillary, *p*=palatinum.

Fig. 33. *Caprolagus (Sylvilagus) Nelsoni* (Merr.) [*Romerolagus Nelsoni*, Merr.]. Palatal view of skull. Nat. size. Popocatepetl, Mexico. Br. Mus. Z. D. No. 97.6.1.5.

Fig. 34. *Lagomys rufescens*, Gray. Palatal view of skull. Nat. size. Br. Mus. Z. D.

Fig. 35. *Caprolagus (Sylvilagus) brasiliensis* (Linn.), juv. Palatal view of skull. Nat. size. Royal College of Science, London.

Fig. 36. *Prolagus sardus* (Wagn.). Palatal view of skull. Nat. size. Pleistocene breccia, Monte San Giovanni (Sardinia). Br. Mus. G. D.

Fig. 37. *Caprolagus (Oryctolagus) crassicaudatus* (Is. Geoffr.). Palatal view of skull. Nat. size. Br. Mus. Z. D. No. 96.6.6.1.

Fig. 38. *Caprolagus (Nesolagus) Netscheri* (Schleg. & Jent.). Palatal view of skull. Nat. size. Br Mus. Z. D.